Design and Manufacturing

Edited by Evren Yasa, Mohsen Mhadhbi and Eleonora Santecchia

Published in London, United Kingdom

IntechOpen

Supporting open minds since 2005

Design and Manufacturing
http://dx.doi.org/10.5772/intechopen.83290
Edited by Evren Yasa, Mohsen Mhadhbi and Eleonora Santecchia

Contributors
Walace Matizamhuka, Tomohiro Sato, Eleonora Santecchia, Mohsen Mhadhbi, Jesper Kristensen, Sayan Ghosh, Yiming Zhang, Waad Subber, Natarajan Kumar, Liping Wang, Genghis Khan, Tojiddin Khayrullaevich Juraev, Kıvılcım Ersoy, Mahmoud Farag, Dina Fouad, Md Enamul Enamul Hoque, Isah Bolaji Kashim, Ebele Erhuanga, Tolulope Lawrence Akinbogun, D. J. Arotupin, O. A. Fatuyi, I. A. Amoo, Kristel Dewulf, Patrick Longhi, Ernesto Limiti, Peer Mohideen Sathikh, Nuno Alves, Cristiana Fernandes, Carla Moura, Paula Pascoal-Faria, Rita Ascenso, Sandra Amado, Berk Barış Çelik, Shiplu Roy Chowdhury, Yogeswaran Lokanathan, Law Jia Xian, Fauzi Mh Busra, Muhammad Dain Yazid, Nadiah Sulaiman, Gargy Lahiry, Murodov Nusrat Murtazoyevich, Naimov Sandjar Tulkunovich

First published in London, United Kingdom, 2020 by IntechOpen
IntechOpen is the global imprint of INTECHOPEN LIMITED, registered in England and Wales, registration number: 11086078, 7th floor, 10 Lower Thames Street, London, EC3R 6AF, United Kingdom
Printed in Croatia

British Library Cataloguing-in-Publication Data
A catalogue record for this book is available from the British Library

Additional hard and PDF copies can be obtained from orders@intechopen.com

Design and Manufacturing
Edited by Evren Yasa, Mohsen Mhadhbi and Eleonora Santecchia
p. cm.
Print ISBN 978-1-78985-865-5
Online ISBN 978-1-78985-866-2
eBook (PDF) ISBN 978-1-83962-889-4

For EU product safety concerns:
IN TECH d.o.o., Prolaz Marije Krucifikse Kozulić 3, 51000 Rijeka, Croatia, info@intechopen.com or visit our website at intechopen.com.

Meet the editors

Dr. Evren Yasa graduated with her degree in Mechanical Engineering from the Istanbul Technical University and completed her master degree at the University of British Columbia on volumetric error modeling and compensation. She received her Ph.D. degree with her thesis on "Combined Process of Selective Laser Melting and Selective Laser Erosion/Laser Re-melting" at the Catholic University of Leuven, and won the "Emerald Outstanding Doctoral Study-Highly commended" award with her doctoral dissertation. After her Ph.D. study, she worked as a senior engineer at TEI, a GE-joint venture company specializing in manufacturing aero-engine parts, where she led Additive Manufacturing projects. Later, she joined Eskisehir Osmangazi University as an assistant professor. Moreover, she has been working as an independent expert in laser-based manufacturing on behalf of European Commission in FP7 and Horizon2020 projects.

Dr. Mohsen Mhadhbi obtained his Ph.D. degree from the Faculty of Sciences of Sfax, Tunisia. He is currently Assistant Professor of Chemistry in the National Institute of Research and Physical-chemical Analysis, Tunisia. His research interests include inorganic chemistry, material engineering, intermetallics, and powder technology. He has published works in national and international impacted journals and books. He is a teacher in inorganic chemistry. He has supervised several researchers in materials science. He is a member of various scientific journals and associations and has been serving as an editorial board member of repute.

Dr. Eleonora Santecchia (PhD) is Assistant Professor in Metallurgy at the Marche Polytechnic University (UNIVPM) located in Ancona, Italy. Her current research activities are mainly focused on metal additive manufacturing and, in particular, on the laser powder bed fusion (LPBF) and direct energy deposition (DED) techniques. She received her Master Degree (cum laude) in Thermomechanical Engineering at the Marche Polytechnic University (Ancona, Italy) in 2010, and obtained her Ph.D. in Materials, Waters and Soils Engineering on March 2014 (Marche Polytechnic University, Ancona, Italy). She accomplished a two years postdoctorate fellowship at Qatar University in Doha (Qatar) working on the Project NPRP 5-423-2-167 "Advanced ultra-hard nanostructured coatings for innovative applications in mechanical and chemical industries". From 2016 to 2019 she worked as an INSTM Postdoctoral Researcher at the Marche Polytechnic University (Ancona), within the European Project DREAM H2020 "Driving up Reliability and Efficiency of Additive Manufacturing". She is experienced in microstructural characterization by scanning electron microscopy, energy dispersive spectroscopy, and X-ray diffraction techniques. Furthermore, Dr. Santecchia is an enthusiastic additive manufacturing researcher, with a particular passion for laser-based 3D printing techniques.

Contents

Preface

Design is a fundamental discipline which regulates all the aspects of science and research, and best practices in industrial environments. Whether referring to it as the basic design of experiment (DoE) or the complex organic shaping of an object, design is the very first step to take in order to accomplish any activity in science, research and industry. The clear definition of the objectives to reach by the end of the programmed activity helps shaping the way to pursue them and to choose the most appropriate options available to boost this process.

This book includes four sections starting form the first one outlining the design perspective applied to different technological case studies (electro-mechanical assemblies in the aerospace industry, multi-objective optimization), addressing also the topic of sustainability in terms of design of biodegradable compounds and in the integration of sustainability in the strategic stages of an innovation process. Section two highlights how design and manufacturing are going hand in hand in the booming technological field of additive manufacturing, which is giving rise to a new era of sintering, allowing to obtain new materials and to process them in a more effective way, involving also rapid physical models to boost industrial design. The third and fourth sections address the implementation of design for manufacturing in the case of agricultural applications with specific case studies (ceramic water filter press for use in Nigeria, and Agriculture Machinery), and for biomedical applications, with a particular focus on the use of additive manufacturing of bioscaffolds for developing tissue-engineered constructs, and additive manufacturing approaches for full bone regeneration.

This book will shed a light on the potential of design and manufacturing and we hope that it will inspire students, researchers and professional to rethink their design and manufacturing approach.

Evren Yasa
Eskisehir Osmangazi University,
Turkey

Mohsen Mhadhbi
National Institute of Research and Physical-Chemical Analysis,
Tunisia

Eleonora Santecchia
Marche Polytechnic University,
Italy

Section 1

Design Perspective

Chapter 1

Design for Manufacturing of Electro-Mechanical Assemblies in the Aerospace Industry

Ernesto Limiti and Patrick E. Longhi

Abstract

Electronic design engineers struggle continuously to obtain a satisfactory trade-off between item performance and cost. On one hand, they would like to employ the best material and components available on the market and opt for time-consuming manufacturing processes in order to obtain high-performance parts. On the other hand, such choice would lead to high recurring cost making the part less attractive in the market. In this scenario, industrial engineering team becomes a crucial industrial entity. It assists the Design Engineers by providing design rules or guidelines. This guidance is intended to provide recommendation to the development team in order to define what is technically feasible and achievable inside an industrial process contest. These rules should not be too strict in order to guarantee acceptable part performance and therefore market attractiveness. The rules contain guidelines on mechanical, process and material aspects. This chapter will focus on design for manufacturing of electro-mechanical parts for the aerospace industry typically being a high-end and high-performance part. Nevertheless, cost and time remain a key aspect to guarantee. The effects of such rules on mechanical and electrical performance will be highlighted and discusses, with a specific focus ion high frequency electrical assemblies (1–30 GHz). It will also contain a review on microelectronic production techniques that impact on the part's electrical performance.

Keywords: design guidelines, design for assembly, design rules, aerospace products, avionics, preferred part list, prototyping, additive manufacturing, continuous improvement, six sigma, product manufacturing figure

1. Introduction

Designing electro-mechanical systems in the aerospace industry is a challenging task for many reasons. First, the programs may last decade, so when the design phase starts the design team must envisage how the product will be sustained and maintained in 20 or 30 years on. Second: reliability is a *must* in this sector. They cannot be taken for granted or worse of all avoided by the design team. Possibly, this is the most important feature a design team should address. All these features are typically summarized in what is defined as quality management system (QMS) that are the company's processes that overlook the design and production phases trying to guarantee the respect of such important requests. All this does not come for free, on the contrary. Guaranteeing and satisfying all these aspects leads to high costs of the engineering and production phase. Nevertheless, engineering and design teams are constantly pressed by the executive board to deliver cost-effective solutions, in time and in-spec.

In this context, the role of industrial engineering teams inside an aerospace company can play a decisive role in delivering the targeted requirements (time-cost-quality).

In order to do so, the industrial engineering team needs to be part of the design team from the beginning, even during offer proposition if needed. Moreover, its requirements, suggestions and strategies must not be seen as secondary or expendable to meet selected electrical or technical specification. On the contrary, if a particular feature needs to be sacrificed during design phase, this should be a technical performance that is not directly requested by the customer or end-user.

During the design flow, industrial engineering can be engaged in two possible ways:

1. In the final design stages to verify that the part designed by the electrical or electronic engineering team fulfills several conditions regarding physical dimensions, materials employed, interconnects, and so on. In practice, the role of the industrial engineering team is to give a "go ahead" or "modify" decision based on the outcome of a specific checklist compilation and know-how of the manufacturing process. In this context the industrial engineering members act as review body rather than participant of the design team. This approach often leads to difficulties when the production of the part ramps-up since some aspects related to manufacturing were overlooked during the design phase.

2. Early on the design stage to recommend manufacturing related views, propose suggestions and identify solutions that would have been probably rejected by a "purely" engineering team.

In essence, design for manufacturing (DFM) is a development & design issue, not a manufacturing topic. "D" stands for design and therefore "DFM" is a design challenge".

The following sections contain indication on how the industrial engineering team can be effective during the design phase (i.e. implementing best practices for DFM) and in the subsequent production phase in order to proactively sustain and improve the manufacturing processes.

The following terms are often referred to in the rest of the chapter:

- Industrial engineering: a team of people, or a better a division of the company, which is constantly involved in both engineering and manufacturing activities. Its essence is to act as the *trait-d'union* between development engineering and production & operations areas. The team is responsible for representing production requirements and needs in the design team, designing the manufacturing work-flow of the part (work-cell and work-cycle design) and sustaining the part during the entire production time

- Producible/Producibility: the attribute of a part that can be manufactured in a given time and cost constraint thorough industrial repeatable processes featuring a level of quality, for example, compliant with ISO9100 standards.

2. Design for manufacturing production technologies and best practices

The design team should treat manufacturing requests and constraints as a requirement in the same way it tackles the technical requirements posed upon the item under development. Therefore, manufacturing aspects require a design strategy and a verification method.

DFM strategies can be summarized as best practices or design rule. In general, rules can be strict and often are associated with the concept of violation and penalty. An alternative way of implementing the process can be obtained by giving guidelines. The latter are less strict and provide a design philosophy rather than giving strict indications.

An important feature of designing and producing parts in the aerospace industry is that large quantities of the same part to be produced are seldom encountered, as occurs in the consumer market or semiconductor industry. Apart for very specific components, for example, transmit/receive modules inside a phased array, most other parts that compose an electro-mechanical system are usually produced in a scale of a few parts per month or even less.

Trade studies are very important in the aerospace industry. They should be carried out at the beginning of the design phase to identify the most viable solution. It is important to emphasize that the Producibility requirements have the same dignity as the electromechanical requirements expressed technical specifications and the team's objective must be to respect ALL requirements, or identify the most balanced solution among a set of proposed viable solutions.

There are multiple ways to implement a project that fulfills the given requirements and conditions. *N* alternatives can be identified (a minimum of three is suggested). The decision on which design solution is" best" can be taken using a radar chart type diagram, as depicted in **Figure 1**, which shows the specific requirements and the constraints to be considered (industrial, growth capability, business opportunities and so on). It is useful to subdivide the requirements and constraints into NEED-TO-HAVE and GOOD-TO-HAVE categories.

A typical case study is here provided with the aid of **Figure 1**. The goal of the team is to design a microwave electromechanical assembly fulfilling some electromechanical requirements listed in technical specification. Moreover, the part shell be produced within a maximum cost figure (expense of components and labor) and the design cycle shall be less than 12 months long.

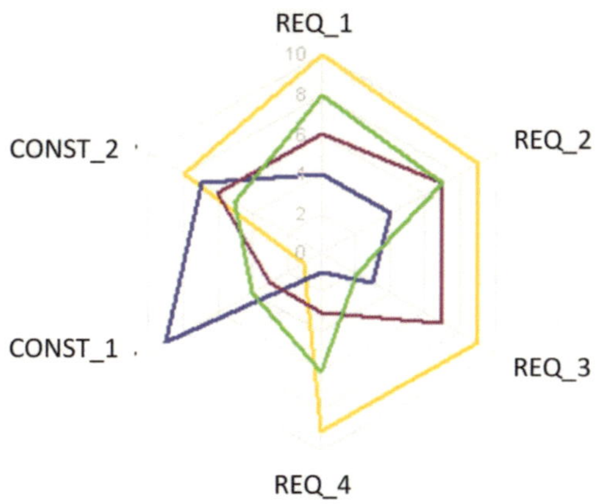

Figure 1.
Radar chart helps understanding design trade options.

Electrical requirements such as gain, noise, signal linearity and DC power consumption can be summarized in REQ_1. Thermo-mechanical requirements, such as maximum temperature of operation and the capability of withstanding certain shocks and accelerations, can be associated to REQ_2. Reliability specifications are considered in REQ_3. The term *space de-rating* is often referred to in aerospace industry to recall the concept that an electrical component can be operated in a sub-optimum electrical condition that reduces the probability of component failure. Finally, cost related aspects are accounted for in REQ_4. Automated assembly greatly influences overall labor cost and shall be also considered when computing REQ_4. On the other hand, automation is a feasible solution when a large quantity of components is to be produced. This is due to the non-recurring expenses associated with the development of automated programs for the specific part. Design time can be considered in CONST_2 while the use of certain component and material to satisfy safety or export prescriptions are considered in CONST_1. An example of safety prescription is REACH requirement applicable in the EU to improve the protection of human health and the environment through the better and earlier identification of the intrinsic properties of chemical substance. The U.S.A. applies a limitation on the export of components only to specific and approved end-user countries (ITAR, ECCN or EAR99).

The yellow line, in **Figure 1**, appears to be a solution featuring high technical merit but requiring the use of some component that is not compliant with safety constraints or export limitations. This is quantified by the low value expressed in CONST_1. On the contrary, the blue line represents a solution that complies with time and material/component prescription but features low technical merit. The green and burgundy curve represent solutions that suitably trade-off between all requirements and constraints. Some requirements may be in contrast against each other. For example, higher electrical performance may be obtained at the expense of poorer reliability or vice versa. Similarly, demanding thermomechanical requirements can be fulfilled if accepting the higher costs of using advanced materials and extra labor time. Moreover, even within the same set of requirements, for example electrical performance expressed as REQ_1 there might be some conflict. Higher gain and linearity is obtained at the expense of greater power consumption.

Typically, the identified solution will cover most of the requirements leaving unsatisfied only a minimal part. Therefore, the best solution is the one having the largest area in conjunction with no points close to the origin of the radar chart, consequently the burgundy curve in **Figure 1**.

The project manager must work to manage the lifetime risk of the product/program linked to the failure to meet these requirements. In the event of conflict, a trade-off must be made between the electromechanical requirements and those of producibility, privileging the latter especially for series production (items with multiplicity ≥5 for one system).

Finally, design guidelines are particularly useful in contexts where most of the assembly is performed manually, whereas rules apply where the process is highly automated and product performance is obtained by-design rather by manufacturing tuning.

2.1 Production technologies and processes

Production of electrical assemblies operating at high frequency requires a set of manufacturing technologies that ranges from packaging to adhesion up to interconnects. The topic is very broad and some aspects are covered in [1]. What is important for this chapter is that several of these processes are manual. While, on one side, manual assembly can help obtain desired product performance on the other it increases tuning time since the "starting point" can be quite far apart due to the larger variability of manual processes. Moreover, at microwave frequencies, interconnects and adhesives influence electrical performance due to the parasitic effects, and therefore must be taken into account during design phase.

A best practice that greatly aids design for manufacturing topics is the manufacturing organization meeting with design engineers to discuss the latest developments in manufacturing technology. Moreover, the Industrial engineering team should periodically provide a report containing investments and improvements foreseen in manufacturing over the following 2–3 years. In this way, the company and the engineering team are well aware of advances in manufacturing and can profitably orient design choices in the future.

2.1.1 Packaging, adhesions and interconnects

Packaging: Kovar is used to match the expansion of alumina or similar ceramics, and is typically used as a carrier for microwave integrated circuit substrates of these materials. If it forms part of the ground plane it is usually plated, or it may be plated to allow soldering or brazing to the ceramic. Kovar is used for small carriers since its density is higher than Aluminum and therefore not advised for large packaging where the overall weight can become too large. On the contrary, Aluminum, thanks to its smaller density is used for the overall packaging.

Adhesion: plays in important role in microelectronics since it provides simultaneously electrical grounding and mechanical bonding. Adhesion at integrated circuit (IC) level can be performed thorough epoxy attach (gluing) or eutectic die attach (brazing). Let us analyze pros and cons of each method.

Eutectic die attach (brazing) is a highly controlled die attach process for high reliability, high accuracy, and high performance devices. To achieve high yield, sophisticated heating and cooling mechanisms are employed. This means controlling that the device heats and cools according to a very strict parameter line. The essence of a eutectic reaction is going from liquid to solid, using eutectic heating and cooling. Eutectic alloys for soldering are composed of Sn (tin), Pb (lead), Ag (silver) and Au (gold). When different metals are combined into alloys, a range of melting temperatures are created with varying proportions of each metal used: AuSi@363°C, AuSn@280°C. The advantage is a very high conductive (thermal and electrical) adhesion obtained at the expense of a manual and very complicate processes (a few seconds or degrees difference in the brazing oven could mean success or failure of the process). **Table 1** reports key attributes of alloys for brazing

Alloy Family	Features	Composition	Melt temp. [°C]
SnPb	Typically used in surface mount assembly. High bond reliability.	$Sn_{63}Pb_{36.7}Sb_{0.3}$	183
		$Sn_{60}Pb_{39.7}Sb_{0.3}$	183–188
		$Sn_{62}Pb_{36}Ag_2$	179
In	Elastic interconnect	In_{100}	156.7
		$In_{50}Pb_{50}$	180–209
AuSn	Strong bond strength. Excellent thermal and electrical conductivity.	$Au_{80}Sn_{20}$	280

Table 1.
Attributes of several alloys for brazing.

microelectronic parts. Important parameters to drive the choice in microelectronic components are the electro & thermal conductivity (to determine in-package device electro-thermal performance) and melt temperature (that implies manufacturing complexity). Gold-Tin alloys (Au/Sn) are typically employed in assembly of microwave devices while Tin-Lead (Sn/Pb) is preferred for the production of digital boards.

Tin/Lead (Sn/Pb) based alloys are the most commonly used alloys for welding on copper, nickel or silver surfaces. The addition (optional) of a small percentage of antimony prevents the transformation of the tin (beta) phase into a tin (alpha) phase called "tin plague", with a reduction in the volume of the alloy mass and a drastic decrease in the mechanical strength of the welded joint. Silver is added to allow soldering on silver surfaces without causing the alloy to over-dissolve the plating metal. All tin-based alloys are strongly discouraged for welding gold surfaces, due to the rapid dissolution of gold in the alloy (scavenging).

Indium-based alloys are particularly useful due to their great ductility, which attenuates or eliminates failure problems resulting from fatigue failure of welded joints, and by the lower solubility of gold in such alloys. About 1% of gold must dissolve in an indium/lead based alloy before the AuIn2 solid phase can be formed, which is stable in equilibrium with lead up to 319°C and acts as a barrier, limiting the further dissolution of gold: a thin film of gold can withstand for 15 minutes in an In50Pb50 alloy bath.

Gold/Tin (Au/Sn) alloy is specifically used to weld gold surfaces without having to use flux, due to the high gold content it contains. It is normally sufficient to use a nitrogen-based inert atmosphere during the process. This alloy is able to dissolve gold in considerable proportions (up to 1–2 microns in thickness) during a normal welding cycle lasting a few minutes, which requires that the surfaces to be gilded have a thicker plating, i.e. at least 3–4 μm.

Epoxy attach (gluing), on the other hand, is a far more easier manufacturing process than brazing. It can be very often automated and the time constraints/temperature constraints of the process are much less critical than brazing. Usually the devices is cured for 30 minutes inside a curing oven at 120°C. Nowadays, silver-loaded epoxy adhesive with high thermal and electrical conductivity are available whose electrical and thermal performance are not far from the ones obtainable with chip brazing.

Interconnects: are the electrical connections between semiconductor devices such as integrated circuits or transistors and the first level of packaging. The most familiar and widely used First-Level Interconnect (FLI) is the wire bond. Wire bonds are available in several types, such as ball bonds, wedge bonds, and ribbon bonds, each with unique variations. The typical wire bond for high-end applications is a wire bond is formed using a gold wire that is typically 25 μm diameter, though high-volume commercial systems at lower frequencies use aluminum wire bonds with diameters as large as 54 μm.

The purpose of the wire bond is to create an electrical connection between an IC and some type of conductor, typically a metal trace. At lower frequencies the wire bond performs as a simple electrical contact between points and is specified at a maximum current handling. However, as frequency increases, wire bonds begin to perform as inductors. The requirements on the wire bond increase as frequency is increased. Typically, the length of the wire is limited to reduce inductance. Also, the shape of the wire bond is specified and in some cases manual accomplishment becomes unavoidable. **Figure 2** depicts the equivalent electric circuit and the corresponding parasitic reactance and resistance as a function of frequency of a 1 mm/25 μm diameter wire bond. As frequency increases, the parasitic effects

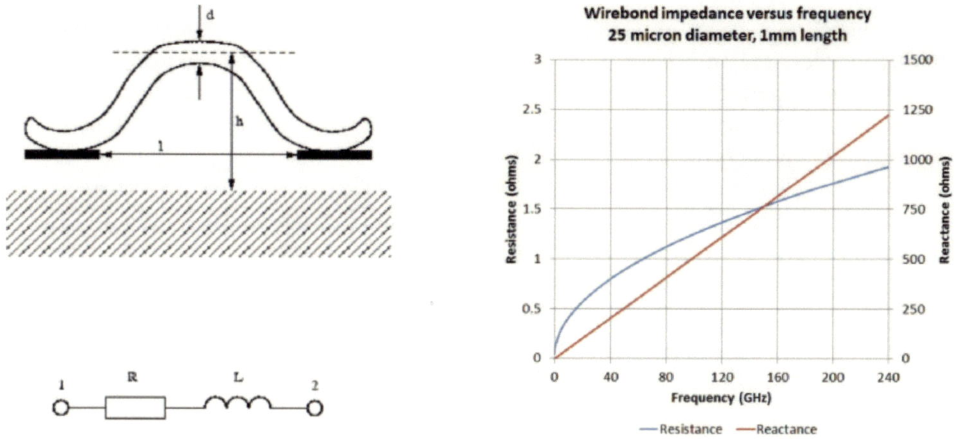

Figure 2.
Bond wire simplified geometry and equivalent circuit (left) and impedance vs. frequency (right).

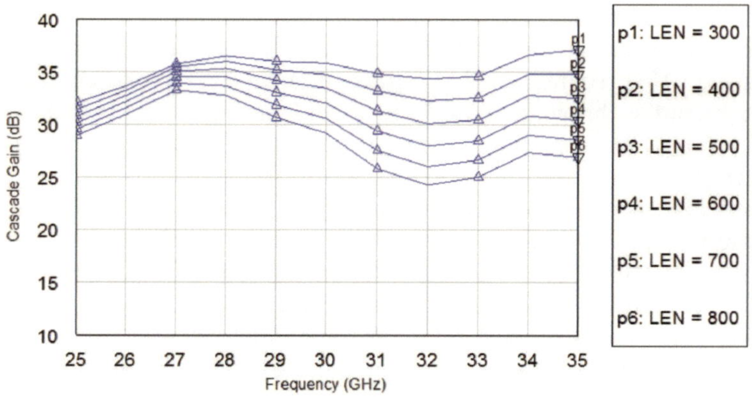

Figure 3.
Bond wire length effect on a RF chain around 30 GHz.

become large and can be compensated only by decreasing wire length, and sometimes operator skill becomes mandatory.

The effect of wire length, and therefore inductance, on a high frequency circuit is demonstrated in **Figure 3**. A *simple* RF chain, composed by a cascade of two amplifying stages, is considered. The two amplifiers, in Monolithic Microwave Integrated Circuit (MMIC) technology, are connected to rest of the circuit trough a pair of wires at the I/O ports respectively. The length of each wire is controlled by a variable "LEN" and is swept from 300 to 800 μm.

The gain is rather flat for LEN = 300 μm (highest curve, marker P1), while it becomes quite rippled and gain drops for LEN = 800 μm (lowest curve, marker P6). Consequently, length of bond wires should be carefully controlled. Occasionally operator ability is essential to obtain the desired electrical performance.

Wire bonds can be connected using ultrasonic bonding, thermos-compression bonding, and thermosonic bonding [2]. Ultrasonic bonding uses pressure and ultrasonic vibrations from a bonding tool to create the bond between the wire and the metal surface. Thermo-compression uses pressure from the bonding tool and high temperature to create the bond. Thermosonic bonding combines ultrasonic and thermos-compression methods to create the bonds.

2.1.2 Automatic vs. manual manufacturing

The choice of manual or automatic assembly is driven by a some parameters. First is the electrical and thermal requirements. In some cases, the requirements could be so stringent that only a manual process is capable of performing a very fine-tuning. For example, when temperature and heat dissipation are critical, then brazing can become the only acceptable solution. The effect of interconnect parasitic were also discussed, in the previous Section 2.1.1, and how operator support can become decisive to obtain acceptable performance, especially at GHz frequencies.

Another parameter to be accounted for is the number of parts to be produced in 1 week, 1 month or 1 year. This number plays a crucial role. If a mass production is foreseen, then manual assembly is not advised due to the lengthy and costly process associated with it. On the contrary, when very few parts are to be produced then manual process is acceptable, also because automatic assembly requires the development of programs and codes with the consequent Non Recurring Expenses (NRE) for developing them.

2.1.3 Additive manufacturing in the aerospace sector

The paradigm of design for manufacturing can be found in Additive Manufacturing (AM) technology. AM represents a key example where an advancement in production technologies enables new engineering concepts that can come to life *only* with this technology. In this sense, it is quintessentially a design *enabled by* manufacturing.

In the aerospace sector, AM is applied mostly on metallic parts (Aluminum, Steel, Titanium and related alloys) rather than composites (plastics) as occurs in the consumer industry. In fact, the initial investment in terms of machinery and training is very high and must be carefully accounted for in the business model.

AM in aerospace has been happening for some time now with many applications, covering everything from the creation of aircraft or helicopter parts, making lighter and more efficient engines, 3D printed turbines etc. 3D technologies generally save on time, money and create stronger, lighter, and more efficient finished products [3].

An example of AM technology and process applied to the aerospace industry is shown in **Figure 4**.

The part itself is not very complex, but is proves how AM can be gainfully exploited to create lighter or more complex structures than the ones previously realized with “prior” technologies.

One of the challenges of the market is the restriction of the volume of construction and the size of the product. An aircraft is made up of very large components and additive manufacturing is today limited to the volume offered by the 3D printer. Most technologies offer solutions with limited print volume, making 3D printing applicable only to small components. So, this constraint that could slow down the growth of the market. Even if so, today’s 3D technologies have already made it possible to create and qualify fairly large (approx. 30 cm) components for space [3, 4] and aviation [5]. Finally, the latest available machines (SLM500, Concept Laser Xline 20000R, EOS M 400) are capable of building even larger pieces.

2.2 Design rules and design guidelines

Design rules can be seen as a set of physical, geometrical, chemical, mechanical limitations. They are very useful when the manufacturing process is constant and

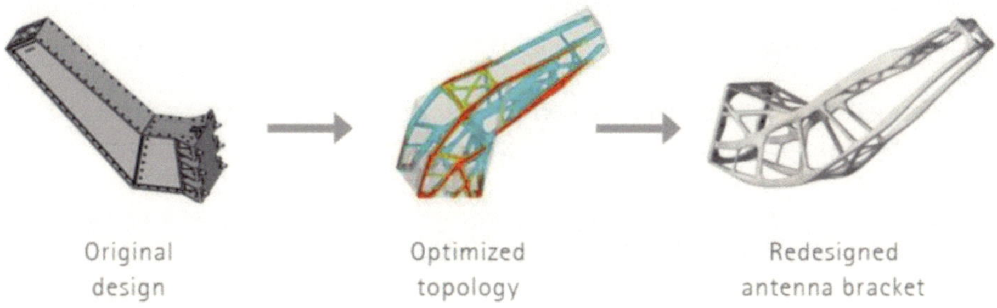

Figure 4.
Metallic part optimization thanks to AM.

repetitive as happens in the semiconductor industries or in large scale production. This paradigm however is less stringent in the aerospace industry since there is not a mass production of items, but on the contrary, a production of a large quantity of different parts each one characterized by very small multiplicity. Moreover, while digital board assemblies can follow rules developed for the consumer market, high frequency microwave assemblies (operating at 100 MHz–30 GHz) are typical of the aerospace industry and suffer from less standardization. Consequently, for the latter guidelines rather than rules should be applied.

Anyhow, rules and guidelines should address the following features that are critical in any industrial manufacturing process:

1. Designing parts for "modularity", i.e. a module is a self-contained component that is equipped with standard interfaces that allow it to be integrated into a larger system. Modularity has several benefits: the product is easy to assemble/re-assemble and most of all, in complex systems, it aids to detect quality problems or non-conformities

2. Designing parts to compensate for process statistics and yield, component and material deviations

3. Ensure the product can be assembled and manufactured using standards processes, i.e. identifiable and written in a production document or drawing without requiring ultra-specialized capabilities or different production approaches for each realized component.

2.2.1 Design rules

Design rules are written to suit a specific production technology. In the electronics for aerospace industry important production technologies are microwave modules and digital boards.

Digital board production uses rules similar to the ones developed for consumer and telecom products, always taking into account that aerospace industry produces a relatively small amount of high-performance products as opposed to consumer market. Anyhow, well known standards can be applied, for example the IPC-2291 "Design Guideline for Printed Electronics" or IPC-2252 "Design Guide for RF/Microwave Circuit Boards" considering class 3 for the aerospace industry.

On the other hand, production of complex microwave parts is very typical to the aerospace & defense sector and seldom finds application elsewhere. This is related to the high cost involved in development and production. Design rules for these objects often end up as a few set of geometrical rules. An example of design rules applicable to hybrid microwave modules or hybrid microwave integrated circuit is given in the following:

- Package dimension not to exceed a certain value so that the part can be manufactured using automatic assembly machines
- Minimum distance between adjacent components, so the part can be assembled using automatic *pick'n'place* machinery.
- Maximum dimension of materials and substrates to avoid cracking due to thermal expansion/compression
- Metallisation and finishing of surfaces
- Geometrical rules regarding thickness, angles, corner radius, shapes, etc.

2.2.2 Design guidelines

Design guidelines provide indications on how to deliver a *design for manufacturing* solution, rather than giving rules and consequently a PASS/FAIL decision. They can be seen as indications that the Design team has to follow in order to design a producible part.

While rules provide a PASS/FAIL criteria, often regarding geometrical or mechanical properties, guidelines provide assort of "sensible" indications so that the design has a higher probability of success. In other words, if the guidelines are followed, very limited manufacturing issues are expected later on. On the contrary, if the design team decides not to follow the guidelines, plenty manufacturing issues during the production stage should be expected.

A typical design guideline could be to avoid overcomplicating the electrical schematic, eliminating unnecessary components. Every component placed inside the schematic should answer to at least one design goal (typically performance, testability or reliability). If a component does not contribute to at least one of these "high-level" design goals, the engineering team should substantiate the reason for which it has inserted. Boothroyd and Dewhurst [6] suggests, among other topics, that unnecessary parts are those that answer "NO" to the following questions:

1. Does the part move relative to other parts in normal operating condition of product?
2. Is it necessary that the part is made of different materials or isolated from other parts such as electrical insulation, heat insulation, or vibration reduction?
3. Does the part have to be isolated from other parts otherwise it is impossible to assemble the products?

If the answer to all questions is "NO", the part is unnecessary and can be integrated with other parts.

Another guideline could be to design parts so that final performance can be obtained after tuning or programming performed in reasonable time and most of all avoid using components (or electrical schematic) so that the overall module performance resides on a specific component of the module. In this case any shortcoming of the component will affect one-to-one the module' behavior.

2.2.3 Prototyping, virtual or real (fast)?

Design engineering team, during the initial design stages, would like to have an initial prototype to test the idea and verify in-lab any limitations that commercial CAD simulations or analysis are unable to predict.

Basically, there two types of prototyping techniques: virtual or real (fast).

Virtual prototyping relies on very accurate model-based CAD simulations. The models are often validated through a previous trial-error-correct cycle. The method is relatively inexpensive, can be very fast and deliver accurate results providing the model itself is accurate.

Additive manufacturing technologies (metal and plastic) provide fast turn-around time to realize real and fast breadboards. In this case, the prototype is real, the time constraints are guaranteed but the exercise can be expensive, compared to virtual.

The choice between real or virtual prototyping can be performed by analyzing the following parameters:

1. Virtual model accuracy

2. Available time and budget constraints

3. Associated Risk mitigation

If parameters 1 and 2 have higher weight then virtual prototyping appears to be the appropriate solution. On the contrary, if design uncertainties are high and risk mitigation is necessary, then real prototyping becomes useful.

2.3 Preferred part list

The objective of a preferred part list (PPL) is to direct the user toward a limited number of component types, covering all design applications. The aim is to avoid duplication and achieve cost reduction and procurement effectiveness [7].

Consequently, you should identify a subset of typically used components to generate your custom PPL. Components belonging to the PPL should be employed "by default", and any derogation from the list should be clearly explained and technically justified.

Definition, creation and sustainment of a PPL should be a company-funded activity and the client-related programs receive the benefit. Like any other engineering effort, the more work put in the initial stages, the less work is required on final stage.

Initial cost is only one consideration for the PPL and is compensated by the value gained over the lifetime of the product (procurement, production and maintenance). Since the cost of introducing d sustaining a PPL in a company is rather relevant at the beginning such choice must be willingly enforced and sponsored by the company's top management (director general end director of engineering). Moreover, the director of the purchasing department has to be actively involved, since he might be tempted, over a short-term period, to prefer cheaper or readily-available parts as an alternative to the parts in PPL.

Components shall be introduced in PPL after analyzing the criteria listed in the following.

Performance history: actual field experience or extensive relevant testing.

- Accessibility: parts that can be purchased from multiple sources, (vendors or/ and distributors).

- Alternating source: same form, fit, and function for parts, but different manufacturers' names and part numbers. (Different manufacturers' crossover part numbers must be equal.)

- Regulatory compliance: RoHS/REACH.

- Reliability figures: mean time to failures (MTTF) or mean time between failures (MTBF).

- Screening: favor pre-screened or tested parts.

- Life span: favor parts with higher Shelf life.

- Economic order or lot quantity: Consider minimum buy.

- Lead-time: consider cost vs. the desired lead time trade-off.

- Bring the strategic suppliers on board the PPL project

Considering the main stages in the product's life-cycle (from concept to maintenance), the possible savings in each phase are examined:

- Research and design: Excluding a newly introduced component's unknown performance will accelerate design validation and testing efforts. Shorter development cycles realized through less component failure issues and time taken for trouble-shooting and reworking breadboards and prototypes. Quicker proof-of-concept results. Parts used from PPL are more likely to be available, and small development quantities can be ready at-hand.

- Purchasing: Material planning is more stable making part procurement less challenging. Strategic suppliers are encouraged if they is actively involved in the company's PPL project.

- Manufacturing: Less line failures using proven parts. Assembly personnel already very familiar with part handling requirements and issues.

- Customer Support: Fewer returns and higher reliability. Practice with frequently used parts promotes a deeper understanding of part behavior and common failure mode and symptom identification. Customer satisfaction with longer life product and fewer returns, and fast turn-around time in repair.

The design engineer who selects the components must choose as many parts as possible from the PPL. Ideally >80% of the bill-of-materials (BOM). By selecting even a majority of the parts from the PPL, the benefits realized from the arguments presented above should be sufficient to encourage the company to validate and enforce the practice of using a PPL.

Finally, it is obvious that the PPL should be created and managed by the Industrial engineering people who are the stakeholders of the activity. In fact, PPL has a n impact on all phases of the product life-cycle. The size of PPL depends on the complexity of the typical system the company develops. For an aerospace company that designs and manufacture avionic systems (radars, electronic warfare, satellite payloads) the size of PPL could be around 2000–3000 components.

2.4 Design for reliability, maintenance and test

As stated many times previously, aerospace products feature high system complexity, and must provide high-performance to be delivered over time and in harsh environment and operating conditions. Consequently, the design team must take into account these aspects when designing the product. Design for Reliability, Maintenance and Test (RMT) is often referred to as design for RMT as if it were a single topic. However, different strategies are employed as clarified in the following to separately guarantee the three topics.

2.4.1 Design for test

Design for test is a crucial aspect to guarantee the part can be efficiently produced during its life-cycle- The part must be designed so that it's key features and characteristics are accessible and verifiable during production test. Keep in mind that in the aerospace industry, practically 100% of the realized parts are fully tested, often over temperature and in mechanically stressful condition (vibration or similar), to verify they are fully compliant to specification and free from manufacturing defects. Moreover, the test is functional and not merely structural. Manufacturing functional tests are carried out to verify that the part is working and function as expected and not just assembled correctly. Functional test on 100% realized HW parts is typical of the aerospace industry to guarantee performance and reliability of manufactured parts and is less applicable to consumer products due to the very high time and cost involved in these kind of test. Finally, aerospace modules that fail the first manufacturing test need to be analyzed and tuned so the part meets the technical specification. Given the time and cost involved in the assembly process, it is illogical that the part should be discarded if the first production test fails. Consequently, designing parts for testability greatly aids the troubleshooting phase, ensuring production people can speedily identify the shortcoming and restore the part.

Given this scenario, it is mandatory that the design team keeps into account these aspects when designing the part. The principle is to add components and interfaces to make it easier to develop and apply manufacturing tests to the designed hardware. At the same time, test engineering department should be consulted in the design phase, so they can bring provide advice and most of all start designing the Automated Test Equipment (ATE) that will be used in production phase but could also be used by the engineering team for product verification and validation. The idea underlying design for test is: Pay less now and pay more later without DFT.

2.4.2 Design for reliability and maintenance

Design for reliability is crucial aspect in the aerospace industry, where reliability is a must considering the mission criticality of these systems [8]. Reliability somewhat depends on the assembly process employed. One indication is to avoid those manufacturing processes that are less repeatable or controllable.

Design for maintenance shared some requirements with design for test, since any maintenance activity starts with identifying the part in failure within the system. Other aspects consist in the designing the parts in a modular way so any failed item can be easily replaced without having replace the entire system or sub-system,

3. Production sustainment

Information and guidelines were provided in the previous section so industrial engineering can proactively contribute in the design team giving correct priority to manufacturing requests. While this activity strongly mitigates manufacturing risks in production stage it does not totally eliminate risks and therefore some process needs to be applied also during product manufacturing life-cycle.

Open literature refers to these processes in many ways: lean manufacturing, six sigma, continuous improvement, kaizen methods, PDCA cycle, and so on [9–11]. Each method has its uniqueness but, fundamentally, they consist in constant proactive monitoring of the manufacturing process to identify deviations in early stage, introduce improvements, observe the expected result and, if the outcome is positive, standardize the new method.

3.1 Continuous improvement and associated methods

Continuous improvement can be obtained by recurrently applying the PDCA cycle to those product and process that demonstrate an intolerable defect rate or more generally deviate from the desired quality/cost/time target.

PDCA cycle consist in performing four steps as graphically visualized in **Figure 5**.

The first step (*plan*) consist in clearly defining what is the "problem" and consequently the expected result (objective) at the end of the process. The Pareto principle can be applied in order to prioritize the (unfortunately) many issues that might be occurring in Manufacturing.

The second step (*do*) is possibly the most challenging for the industrial Engineer. The goal of this second step is to identify the Root causes that prevent the product/process being on-time, on-cost and in-quality. Many problem-solving techniques can be applied. An example shown here is the Ishikawa diagram **Figure 6** that can be very useful since it helps clustering into smaller sub-problems, which become more easily addressable.

Ishikawa "fish bone" diagram method consists in analyzing all pertinent areas and sub-areas of a typical manufacturing process. When a quality issue arises, the industrial engineering team is notified in order to identify the root-cause of the issue and consequently propose a corrective action. This is not a simple task since there are many areas and factors to be investigated. Moreover, some of the production processes and materials may come from tier 1 suppliers and therefore occur outside the company.

Common production issues in the aerospace industry occur when information related to a specific production process is not fully written but relies on the skill of advanced operators. Therefore, a strong practice is to provide very detailed assembly instructions so that lesser skilled operators can produce the part in high quality standards.

Some issues may sometimes occur when the purchasing department, to obtain cost saving, procures a component or a material from a different supplier claiming it is equivalent form, fit and function (FFF). Rarely this is a painless change since there are always some small differences between two components identified as equivalent FFF on to the other.

Environment parameters (temperature and humidity) are rarely a cause of manufacturing deviations since the assembly process is typically carried out in clean rooms or at the least humidity/temperature controlled areas. In the aerospace industry, final assembly is performed in the company while lower level components

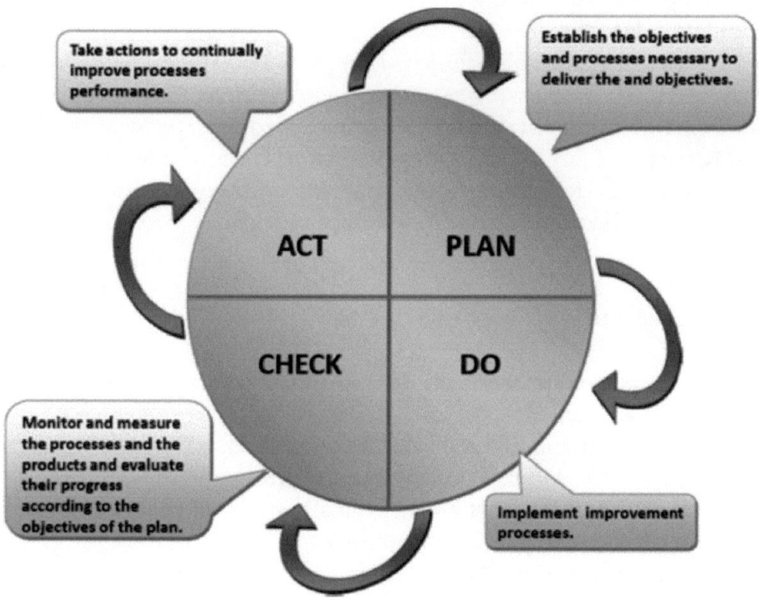

Figure 5.
Plan-do-check-act cycle.

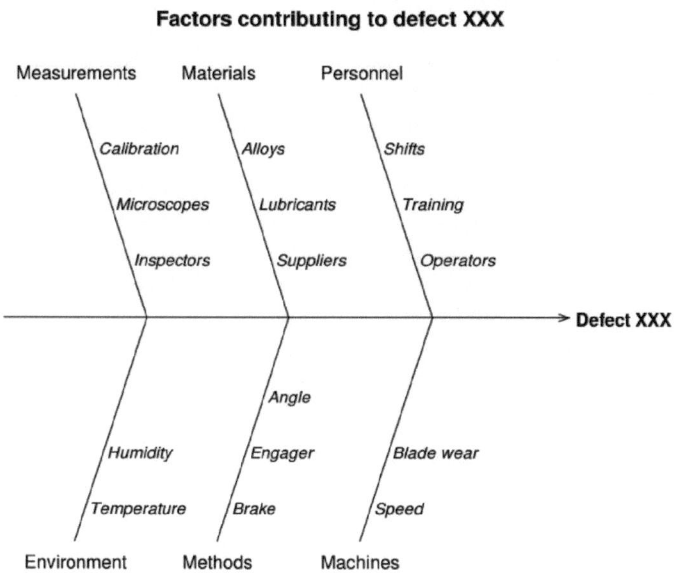

Figure 6.
Ishikawa "fish bone" diagram useful for problem solving.

and sub-systems may be procured from an external contractor. The same holds for some non-critical services that are occasionally outsourced. Consequently, in some cases, the investigation needs to be performed at tier 1 supplier level too in order to investigate and identify the root cause of the problem.

The final two steps are *check* and *act*. In practice, what has been identified and proposed in the previous two stages needs to validated and standardized. In all steps, it is important to be un-biased and all problem inputs should be data-driven.

In this context, systems for tracing non-conformities are vital so the create an effective and populated database.

W. Edwards Deming's famous quote is therefore a cornerstone of this problem solving technique: *"Without data you're just another with an opinion"*.

Another practice that contributes to improve product/process performance are manufacturing and engineering organizations periodically reviewing quality non conformities to determine if engineering changes are required. Creating dedicated interdisciplinary teams to perform a specific improvement project is also useful.

3.2 Product manufacturing sheet and figure

As stated previously, all process/product monitoring and the consequent PDCA cycle should be data driven.

A Product Manufacturing Sheet is useful from which a Product Manufacturing Figure (PMF) can be calculated. The sheet and figure are living documents and figures, in the sense that they must be periodically updated to monitor the improvement of a certain production product/process.

The Product Manufacturing Sheet contains structured information regarding its three macro-topics: design, manufacturing and purchasing.

- DDP (design data package): specification, engineering drawings, data libraries, SW code, design rationale documentation, test planes, are available. List of major engineering changes ongoing, if any.

- MNFR (Manufacturing and workmanship): are all the Tooling/machinery available? Personnel has been trained for the specific product? Automatic test equipment – if necessary – is available? Screening procedures are in place?

- SC (SUPPLY CHAIN) quantifies on-time and on-quality purchasing of the major "buy" items that constitute the product, any obsolescence, vendor rating of the key components.

The Product Manufacturing Figure (PMF) is calculated, as indicated in Eq. (1), by summing the three previously mentioned factors, each having a weight (α, β, and γ) proportional to the importance the company gives to each factor.

$$PMF = \alpha * MNFR + \beta * DDP + \gamma * SC \ldots \alpha = \frac{1}{2}, \beta = \frac{1}{4}, \gamma = \frac{1}{4} \qquad (1)$$

The PMF is computed in the following way:

1. At first, the weight is set for each e parameter (the sum of the weights must be unitary). In Eq. (1), for example, $\alpha = 0.5$, $\beta = 0.25$, and $\gamma = 0.25$. These weights shall remain constant all over the production process.

2. A figure between 0 and 100%, according to a checklist, is computed for each parameter (MNFR, DDP and SC) in Eq. (1). This figure changes in time as the three topics improve (or worsen). Checklists become handy to substantiate the figure—between 0 and 100%—associated to each parameter. Moreover, Quality Notifications can be used to obtain useful information of product non-conformities.

3. Consequently PMF is calculated.

PMF close to 100% indicates the part can be fully produced on-time, in-spec and on-quality. Lower values indicate that you should expect some contained derogation of one of the three parameters. PMF < 40% indicates that the product is not enough for mature for an Industrial-grade production and important improvements have to be applied to one or more of the three parameters. Furthermore, PMF is a *living* index, since it can be computed periodically to register changes in the three parameters. For example, MNFR could improve after a set of tooling is made available or SC worsen if a component becomes obsolete.

4. Conclusions

Evidence so the Industrial engineering team can proactively contribute to designing parts and address manufacturing issues during the design follow is provided. In this chapter, the starting point is deep knowledge and understanding of the critical technologies that apply to each manufacturing process and their impact on product assembly and performance. Once the technologies have been considered, the key-points Industrial engineering team must engage are: involvement from the early stages, definition of rules and guidelines for manufacturing.

Occasionally, the prior activities are not sufficient and some product improvement must be carried out during the production process. Specific continuous improvement activities (PDCA cycle) and also detailed tools and figure to quantify “design quality” in manufacturing have been provided.

Acknowledgements

Patrick E. Longhi would like to thank friends and colleagues at Elettronica Group in Rome (ITA) for the many fruitful and insightful technical discussions during his time spent in the company as a Microwave Design Engineer and Industrial Engineer.

Author details

Ernesto Limiti and Patrick E. Longhi*
Department of Electronic Engineering, University of Roma Tor Vergata, Rome, Italy

*Address all correspondence to: longhi@ing.uniroma2.it

References

[1] Sturdivant R. Microwave and Millimeter-Wave Electronic Packaging. Norwood, MA 02062: Artech House; 2014

[2] Harman G. Wire Bonding 3rd Edition. New York, NY, USA: McGraw-Hill Education; 2010. ISBN-13: 978-0071476232

[3] Froes F, Boye R. Additive Manufacturing for the Aerospace Industry 1st Edition. Amsterdam, The Netherlands: Elsevier; 2019. eBook ISBN: 978-0128140635

[4] Available at: https://www.thalesgroup.com/en/worldwide/space/press-release/thales-alenia-space-takes-3d-printing-series-production

[5] Available at: https://www.boeing.com/features/innovation-quarterly/nov2017/feature-thought-leadership-3d-printing.page

[6] Boothroyd G, Dewhurst P, Knight WA. Product Design for Manufacture and Assembly. Boca Raton FL: CRC Press; 8 December 2010. ISBN: 978-1420089271

[7] ECSS-Q-60-01A, European Preferred Parts List (EPPL). ESA standards

[8] Kailash C. Kapur, Reliability Engineering. NJ: Wiley; 2014. ISBN-13: 978-1118140673

[9] Frank Voehl H, Harrington J, Mignosa C, Charron R. The Lean Six Sigma Black Belt Handbook: Tools and Methods for Process Acceleration. Boca Raton FL: CRC Press; 2013. ISBN: 978-1466554689

[10] Bhat S. Continuous Improvement: 30 Proven Tools to Drive Profitability, Quality and Operational Effectiveness in Manufacturing & Service Industry. 1st edition. Canada: Shifting Paradigms Publications; 15 May 2017. ISBN: 978-0994825506

[11] Stern TV. Leaner Six Sigma: Making Lean Six Sigma Easier and Adaptable to Current Workplaces. Boca Raton FL: CRC Press; 2019. ISBN: 978-1138387928

Chapter 2

Industrial Applications of Intelligent Adaptive Sampling Methods for Multi-Objective Optimization

Jesper Kristensen, Waad Subber, Yiming Zhang, Sayan Ghosh, Natarajan Chennimalai Kumar, Genghis Khan and Liping Wang

Abstract

Multi-objective optimization is an essential component of nearly all engineering design. However, for industrial applications, the design process typically demands running expensive computer code and/or real-world experiments putting the design process at risk of finding suboptimal solutions and/or not meeting budget constraints. As a first step toward a remedy, meta-models are built to mimic the response surface at a much lower query cost. We cover a time-tested technology specifically tailored to limited-data scenarios called Bayesian hybrid modeling (GEBHM) developed and maintained at General Electric (GE) research. GEBHM offers Bayesian mean and principled uncertainty predictions allowing a second technology called intelligent design and analysis of experiments (GE-IDACE/IDACE) to perform the optimization task using an adaptive sampling strategy. This chapter first covers the theoretical framework of both GEBHM and GE-IDACE. Then, the impact of GEBHM/GE-IDACE is demonstrated on multiple real-world engineering applications including additive manufacturing, combustion testing, and computational fluid dynamic design modeling. GEBHM and GE-IDACE are used daily and extensively within GE with huge impact in the form of 30–90% cost reduction and superior engineering designs of competitive products.

Keywords: intelligent design and analysis of computer experiments, GE-IDACE, IDACE, Bayesian hybrid modeling, BHM, GEBHM, GE-BHM, Gaussian process, GP, GE, adaptive sampling, meta-model, surrogate model, uncertainty sampling, Bayesian global optimization, BGO, multi-objective optimization, industrial, industrial design, engineering design, application, real-world, predictive uncertainty, machine learning, ML, artificial intelligence, AI, robust optimization, desirability, expected improvement

1. Introduction

Through better engineering design we can build superior aircraft engines with higher efficiency and lower weight, wind turbines that can produce more energy at a reduced cost, and steam turbines that can reach higher efficiencies, to mention a few examples. The importance of a fast design process which produces global optima with the fewest amount of resources is obviously fundamentally crucial. This process, for the purpose of this chapter, can be thought of as a multi-objective optimization problem. Consider as an example the design of aerodynamic airfoil shapes. First, the performance is produced, followed by a mechanical and aeromechanics assessment. Aeromechanical feedback and reactive aerodynamic redesign rely heavily on the domain expertise of the design engineers. It is not atypical to cycle through 50 of these iterations to obtain a design that satisfies mechanical (stress, creep, fatigue to mention a few) and aeromechanical (say, clean Campbell and flutter resistant) requirements. During these cycles, data from expensive computational codes and/or real-world experiments are collected and the design cycle continues in a direction suggested by this information. Generally speaking, as some examples of the key contributions toward high resource requirements are expensive computer simulations such as computational fluid dynamics (CFD) and ANSYS. In some cases, real-world experiments need to be performed, e.g., when it comes to passing FAA certification.

With this, it should also be clear that engineering design is performed under very strict budgets. Each datum obtained whether from a simulation, physical experiment, or an expert needs to be as informative toward the goals we are trying to accomplish as possible. In some cases, it can take weeks or months to evaluate a single datum. In this case, a meta-model is built on a small representative set of data. This can be Gaussian processes (GPs) [1–5], Bayesian hybrid modeling as used at GE [6, 7], or polynomial chaos expansions (PCEs) [8, 9].

One of the key goals of this chapter is to present the state-of-the-art industrial tools toward achieving the best possible optima under strict budget constraints. Specifically at GE we are regularly seeing a reduction in cost needed to obtain the same level of information on the order of 30–90%. This leaves more room in the budget for finding even better, more competitive, designs as hitherto possible. As a consequence of the technologies covered in this chapter, we are building better aircraft engines, improving our steam turbines, and harvesting more wind energy because of this. There is still a lot more to be invented and improved, but the following sections will give an idea of where we currently stand.

2. Theoretical framework

In order to introduce some nomenclature and to lay the foundation for surrogate modeling, and adaptive sampling, consider a concrete example of an engineering design task.

Before diving into the details of the setup it is worth briefly discussing how the data is collected. We obtain data in this design task either from real-world experiments or from computer experiments the latter defined in Ref. [10]. A computer experiment consists of running an expensive complex computer code for a set of different inputs. One of the main motivations of using computer codes is to approximate and thereby speed up costly real-world experiments in order to reduce the engineering design cycle time.

Continuing the example, imagine designing a wing blade described by a set of M geometric design variables/features/input dimensions. The blade is part of an engine and for a given blade design, i.e., for given values of the M features, the engine produces two outputs/objectives one which is the efficiency and the other being the mass flow rate. The *design space* is an M-dimensional space containing the set of all possible geometries we can consider. In this example, we want to maximize the efficiency and to minimize the mass flow rate but in general, we may have N objectives. When $N>1$, we are perform multi-objective optimization, and later in this chapter, we explore a set of sampling strategies suitable for various values of N. The set of all points in objective space form a *response surface* which is assumed to be extremely costly to exhaustively explore putting the search for global optima at risk. Toward reducing the cost of the overall design process regardless of whether the data is obtained by real-world experiments or complex computer codes, a key ingredient is meta-/surrogate modeling where an approximation to the response surface is obtained by generalizing its behavior under certain assumptions (such as smoothness) based on few observed instances of said surface. This is typically done by querying a set of initial points on which the surrogate model is built. Following this, the meta-model is extremely cheap to query in comparison to the surface itself. These meta-models require, typically, only a handful of data to construct an accurate representation of the response surface. An important question however arises on which points to pick, i.e., on how to form the design [10, 12–14]. Given a budget on the total number of data points, how are new points added to this design sequentially? This is where adaptive sampling comes into play and GEBHM is used in conjunction with a powerful technology called intelligent design and analysis of computer experiments (GE-IDACE/IDACE) [15, 16].

In this section, we provide an overview of the mathematical framework of the GEBHM and GE-IDACE technologies introduced in Section 1. Further theoretical details and application coverage can be found in Refs. [7, 11, 15–26].

2.1 Bayesian hybrid modeling (GEBHM/BHM)

In industry applications, it is not uncommon for the data to be multi-dimensional, noisy, highly non-linear, and expensive to collect. On a day-to-day basis, we address the challenge of enabling a robust and uncertainty certified design utilizing both limited expansive simulations data and noisy field measurements. GE Research has an in-house software framework for advanced Bayesian modeling and machine learning named GEBHM, sometimes we shall refer to this as simply BHM, which enables the combination of multiple numerical simulations and experimental sources of data in one unified workflow. As shown in **Figure 1**, GEBHM capabilities are: uncertainty propagation and quantification, sensitivity analysis, full Bayesian model calibration, meta-modeling, multi-fidelity analysis, and adaptive design of numerical experiments. The theoretical framework of the GEBHM is based on Kennedy O'Hagans approach to modeling and fusing simulation and experimental data with associated uncertainties using GPs [6]. The noisy high-fidelity model is represented as Gaussian process aggregated from a linear combination of a low-fidelity model and a model discrepancy function $\delta(\cdot)$ as

$$y(x_i) = \eta(x_i, \theta) + \delta(x_i) + \epsilon, \quad \text{for } i = 1, \dots, n, \tag{1}$$

where $y(\cdot)$ is the (high-fidelity numerical model or experimentally measured) response. The low-fidelity model is $\eta(\cdot, \cdot)$ and discrepancy term $\delta(\cdot)$ are modeled by separate GPs. The design variable is denoted by x_i, while the calibration parameters are denoted by θ. GEBHM allows for calibration of a set of model parameters in the low-fidelity model. For example, this could be parameters in a CFD simulation that we want to tune/calibrate in order to match real-world experimental runs as closely as possible. The measurement error is denoted by ϵ.

The GP hyperparameters are learned using a Markov Chain Monte Carlo (MCMC) technique based on an Metropolis-Hastings-within-Gibbs algorithm [27, 28] with univariate proposal distributions for the posterior distribution

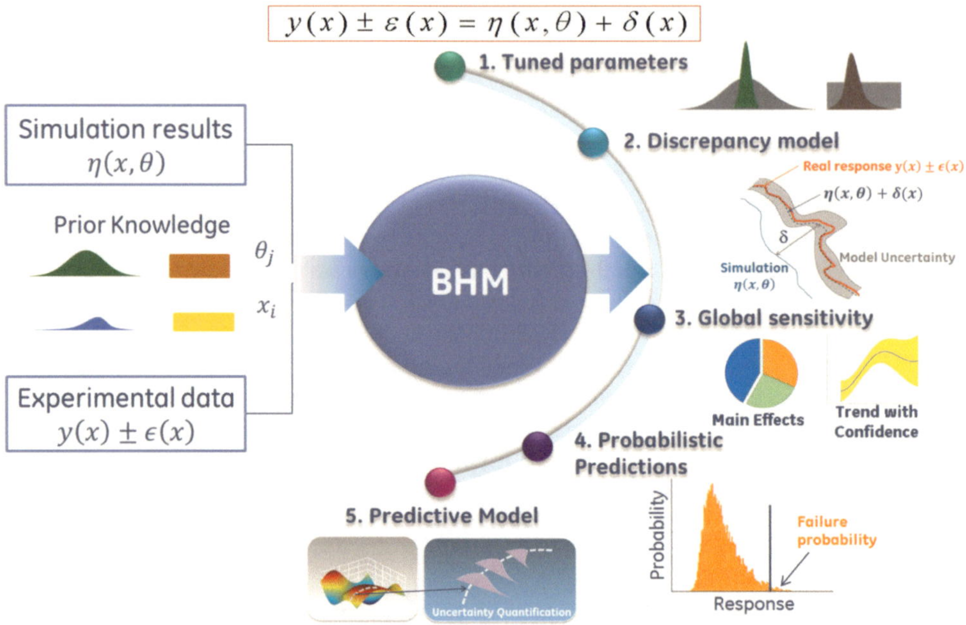

Figure 1.
A diagram of GE's Bayesian hybrid modeling (GEBHM/BHM). From left to right, varying-fidelity-level data (e.g., simulation vs. experimental) can be input to GEBHM. GE research has added a large range of capabilities over the years listed to the right which include parameter tuning, building discrepancy models between low and high fidelity data, informing the designer about which inputs mostly impact the outputs via a global sensitivity analysis, performing probabilistic predictions including tail probabilities, and building high-accuracy predictive surrogate models.

updates. MCMC generally converges toward the most probable values for the parameters which best explain the data [29] from which representative samples can be obtained. To avoid overfitting the high-fidelity data, the initial values of the hyperparameters of covariance matrices are updated with the current realizations at every MCMC step, and realizations from the posterior distributions of the model parameters are produced.

2.2 Intelligent design and analysis of computer experiments (GE-IDACE/IDACE)

While meta-models offer very low query times for response surface values they are still only approximate, especially when built on small datasets. Thus, having the meta-model does not generally mean that we can use this entirely in place of the response surface. However, we can use it as a guide to seek out new locations in the design space that are promising toward our goal which could, e.g., be optimization or to produce the most accurate surrogate model possible. The focus in this chapter is on the prior goal of global optimization.

GE-IDACE, also sometimes referred to as simply IDACE, uses the expected improvement (EI) [30] method to explore and exploit the design space for obtaining the global optimum with the fewest possible resources, see **Figure 2**. Without loss of generality, EI defines the *improvement* of a new design point x as $I(f(x)) = \max\{f_{\min} - f(x), 0\}$, where $f_{\min}$ is the current best point (also called the incumbent) and we will suppress x going forward. The surrogate model predicts a distribution for f denoted p_f. This makes $I(f)$ a random variable. In face of this randomness, we are just interested in knowing how large the improvement is expected to be:

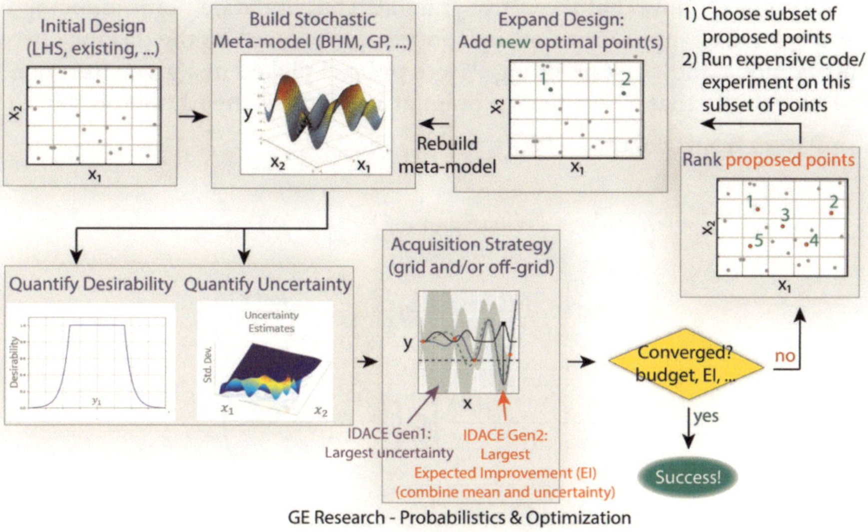

Figure 2.
Diagram of intelligent design and analysis of computer experiments as used by GE research (GE-IDACE/ IDACE). Starting with an initial design, e.g., from LHS, a stochastic model, such as BHM, is built on this. Via the stochastic predictive distribution, the desirability and uncertainty are quantified which combines into an acquisition strategy (such as: Pick points to optimize an objective—that is expected improvement (EI)). Then, we check for convergence and rank a new set of points to be run and then re-build the stochastic model completing one iteration of GE-IDACE. We iterate until convergence defined as either budget exhaustion, the EI, e.g., reaching a specific value, or something else.

$$\mathrm{EI} = \int_{-\infty}^{f_{\min}} (f_{\min} - f) p_f \mathrm{d}f. \tag{2}$$

In each iteration of GE-IDACE, the point with maximum EI is added to the design.

In what follows, we review some GE-IDACE multi-objective optimization EI methods that we have found to work well in practice but emphasize that more research is needed toward getting faster at locating the global optimum in multiple dimensions under increasingly stricter budgets [15].

2.2.1 Multiple objectives: centroid method for two dimensions

Many approaches exist for running IDACE with multiple objectives [31–39]. Here, consider the two-dimensional case and the so-called centroid method which shares a similar intuition as its one-dimensional counterpart Eq. (3). In this methodology, each candidate point from the design space is imagined to create a centroid point, the equations for computing this point are given below. This centroid point, which is located in output space, is then compared to its closest Pareto point on the current frontier [40, 41]. For simplicity, consider now two different candidate/ design points where we compute the associated centroid point for each. Then, for the ith candidate point x^i the centroid point in two-dimensional output space is $\overline{f}(x^i) = \left(\overline{f}_1(x^i), \overline{f}_2(x^i)\right)$, the Pareto point P^i on the current frontier closest to $\overline{f}(x^i)$ is computed and the distance between $\bar{f}(x^i)$ and P^i quantifies the value of adding candidate point x^i. Note that $P^i \neq P^j$ in general. The candidate point inducing the biggest expected change in the Pareto frontier as measured by the distance between its centroid point and corresponding Pareto point is picked in a given iteration.

The probability that a new design point at x improves the ith member of the current Pareto front, denoted $\left(f_1^{*,(i)}, f_2^{*,(i)}\right)$, is:

$$P\left[f_1(x) < f_1^{*,(i)} \cup f_2(x) < f_2^{*,(i)}\right] \equiv P[I]_{2D} = \tag{3}$$

$$\Phi\left(\frac{f_1^{*,(i)} - \mu_1(x)}{\sigma_1(x)}\right) + \Phi\left(\frac{f_2^{*,(i)} - \mu_2(x)}{\sigma_2(x)}\right) \tag{4}$$

$$-\Phi\left(\frac{f_1^{*,(i)} - \mu_1(x)}{\sigma_1(x)}\right)\Phi\left(\frac{f_2^{*,(i)} - \mu_2(x)}{\sigma_2(x)}\right). \tag{5}$$

The two-dimensional EI then becomes:

$$E[I(\boldsymbol{x})] \equiv \mathrm{EI}_{2D} = P[I]_{2D}\sqrt{\left[\bar{f}_1(x) - f_{1,c}^*(x)\right]^2 + \left[\bar{f}_2(x) - f_{2,c}^*(x)\right]^2}, \tag{6}$$

where $\left(f_{1,c}^*(x), f_{2,c}^*(x)\right)$ is the Pareto point, among all Pareto points of the current Pareto frontier, which is closest to the centroid point $\left(\bar{f}_1(x), \bar{f}_2(x)\right)$. The details are available in Ref. [15] where also another method assuming a convex hull Pareto shape is discussed, which in certain scenarios outperforms the centroid method.

2.2.2 Multiple objectives: hypervolume method for any dimensions

The hypervolume EI method is presented to handle high-dimensional objective spaces beyond two [15, 42]. Drawing an analogy with the one-dimensional case, in multi-dimensional objective space, the hypervolume is considered a measure of the current known minimum point (the Pareto front). The difference in the hypervolume between the current Pareto front and the new Pareto front resulting from adding a candidate point is used to define the EI. Accordingly, the EI in multi-dimensional objective-space, $\mathrm{EI}(f)$, gained by adding a new point x with objective values $f(x) \equiv f$ to the design space is defined as

$$\mathrm{EI}(f) = \int [\mathrm{HV}(f^* \cup f) - \mathrm{HV}(f^*)] p_f(f)\ df, \tag{7}$$

where $\mathrm{HV}(f^*)$ denotes the hypervolume contained by the current Pareto front. In general, the expectation integral in the hypervolume EI method is simplified by decomposing it into sub integrals over hyperrectangles, please see Ref. [43]. Further simplification to this integral can be achieved by assuming the predicted output components are independent [44]. To reduce the cost of computing the hyperrectangles integrals, Monte Carlo approximation can be utilized [45] as well as using a suitable merging approach to decompose the integral as presented in [31]. It is worth mentioning that depending on the domination of the proposed point over the current Pareto front, different levels of improvement can be gained [15].

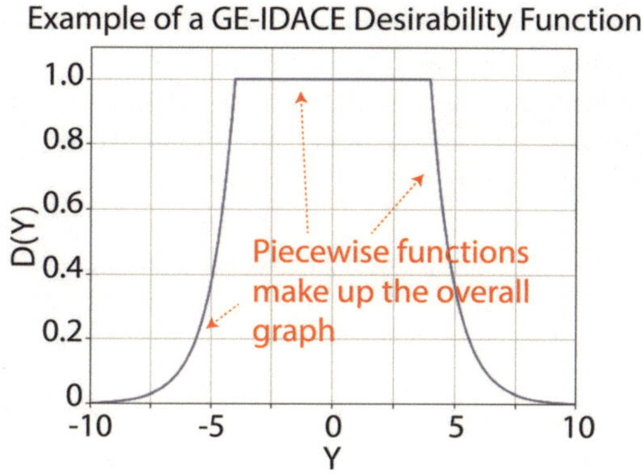

Figure 3.
An example of how the objective space, here one-dimensional, is split into regions (here (−10, −4), (−4, 4), and (4, 10)) and separate independent desirability functions with user-defined parameters (not directly shown) are defined. In this case, candidate points with predicted objective values from the surrogate model in the range (−4, 4) are primarily favored since the y-axis being D(y) takes the largest values. Note that D(y) need not integrate to any specific value (such as unity). Desirability provides a lot of flexibility to GE-IDACE. As a simple example, it enables us to target a specific objective value instead of maximizing/minimizing it.

2.2.3 GE-IDACE with desirability

The physical programming technique can help the engineer to guide the design algorithm toward the desirable regions of the Pareto front [46]. For example, let the ith objective in a multi-objective design problem be denoted by $Y_i \in (-10, 10)$. Then, by dividing the range of possible values into four segments (e.g., $(-10, -5)$, $(-5, 0)$, $(0, 5)$, and $(5, 10)$), the design engineer can assign a desirability for each sub-range as a highly desirable, acceptable, undesirable, and unacceptable. An aggregate desirability function that is formed from these individual ranges is used to rank the Pareto points. Within the GE-IDACE framework, we help the designer to define desirabilities from a set of functions by decomposing the objectives into ranges as shown in **Figure 3**. The desirability function in this case is one-dimensional and decomposed into three regions whereby different desirability functions are assigned by the designer. The figure shows that the candidate points mostly favored have an objective value predicted to be in the range (−4, 4).

Next, we extend the hyperrectangle approach to account for desirability as follows. Specifically, the designer chooses, for each objective, ranges of the objective which are considered highly desirable, acceptable, undesirable, and unacceptable. Closely following the ideas in Ref. [47], consider the following quantity called expected desirability of improvement (EDI):

$$\mathrm{EDI} = \int_{\mathrm{I}(f)>0} D(f) p_f \mathrm{d}f. \tag{8}$$

Given the predictive distribution p_f for some new point, EDI is the mean desirability of the predictions that improve the current Pareto front.

3. Industrial applications of GEBHM/GE-IDACE

Having covered the theoretical framework for both GEBHM and GE-IDACE this section turns to demonstrate their application to real-world engineering applications. We consider first additive manufacturing (AM) which is a vital process for many engineering design applications and is bound to further transform manufacturing. In essence, AM can be defined as the process of overlying layers to create a three-dimensional objects [42, 48]. We will show how GE-IDACE reduces the design cycle time from 6 months to a few weeks.

As a second application, we consider combustion testing where the goal in this case is to maximize load while keeping exhaust and temperature within specific limits. We demonstrate that GE-IDACE can help guide the test into regions with 20% more points from critical areas compared to status quo.

Finally, we demonstrate how well GE-IDACE does for expensive complex computer codes such as CFD modeling. We show that GEBHM/GE-IDACE helps reduce the number of test points by a factor of three when compared with neural network modeling and optimization.

3.1 Additive manufacturing

As a main example of AM applications utilizing GE-IDACE we consider Direct Metal Laser Melting (DMLM) but mention that GE-IDACE is also used for feature-based qualification methods for Directed Energy Deposition having a big positive impact.

DMLM is a key modality of additive manufacturing that focuses on 3D printing of metallic materials. Printing metals is in itself a complicated task due to the microstructural instabilities from melting of the metallic powder. It is especially complicated for superalloys since as-built parts from DMLM are highly prone to microcracking and other microstructural deficiencies. So it is of primary importance to identify what the processing parameters are for the hard-to-process Nickel-based superalloys, and that process has been proven to be non-trivial. The lead times for processing parameter development for these types of alloys are typically on the order of several weeks to months, which means increased cost and the inability to introduce new materials in the additive marketplace. In order to reduce the cycle time when developing the processing parameters for DMLM for hard-to-process alloys, we have extensively utilized GEBHM and GE-IDACE to collect data in an intelligent manner [49]. Typically the key parameters that dictate the processing of additive parts are the laser power, the laser speed, etc. GE-IDACE automated the process toward obtaining design points, i.e., processing parameter combinations, which were most informative to the model and which would guide us in the direction of the optimal solution(s). We used quantified characteristics of the microstructural deficiencies as our outputs/objectives. Parts were built in the additive machine and then characterized by sectioning the parts, imaging the sections, and performing automated image analysis. This enabled us to analyze a large number of images extracting specific defect information such as porosity, keyholes due to unmelted powder, etc. We are interested in porosity because it affects the mechanical properties like yield strength and fatigue life adversely. The GE-IDACE process then constructed a model of the output microstructural defects as function of the input process parameters. We utilized both variance minimization (uncertainty sampling) and EI-based optimization to exploit and explore the design space to identify the optimal solutions faster. Using GE-IDACE with GEBHM, we were able to reduce the cycle time in identifying the optimal process parameter window for a superalloy from 6 months to a few weeks.

Figure 4 shows the progression of the collection of data (only two dimensions shown in a multi-dimensional problem). We can clearly see that the GE-IDACE methodology helps us explore the design space initially and then start to exploit the optimal solutions in the later iterations to quickly converge on an optimal

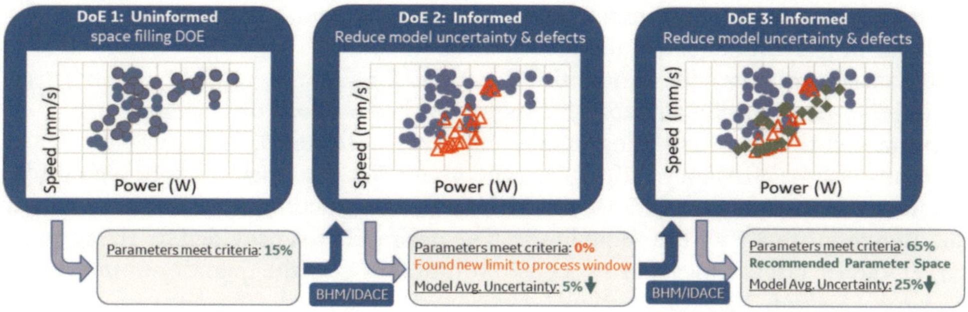

Figure 4.
An example of GE Bayesian hybrid model (BHM/GEBHM)/GE intelligent design and analysis computer experiment (IDACE/GE-IDACE)-based process parameter optimization for a hard-to-process superalloy. The plot on the left shows an initial design from a space-filling and uninformed design of experiment (DoE/DOE). Defects in an as-built part was measured after each DoE. The red triangles in the middle plot are suggested by GE-IDACE based on a GEBHM model built on the blue-circle dataset. As noted in the text box below the plot, DoE 2 reduced the model uncertainty by 5% but did not suggest any datapoints that meet the target defect criteria (not specified here). In DoE 3 on the far right, the green points are suggested by GE-IDACE based on a GEBHM model built on the blue circles and red triangles. By adding informative data, we have added more information to GE-IDACE through the underlying GEBHM models. As a result, at DoE 3 we saw further reduction in model uncertainty (close to 25%) and also excitingly identified a parameter space window where we obtained more than 65% of datapoints satisfying the defect criteria. The figure overall aims to demonstrate the power of the GE-IDACE methodology for performing experimental design.

processing window. This shows that by the third iterations GE-IDACE suggested to us almost 65% of points that satisfy our objectives in defects, while also bringing the overall model uncertainty down. Currently, we are working on expanding this methodology into more complicated structures and additional quantities of interest (QoIs) such as mechanical properties, durability, surface finish etc.

3.2 Combustion testing

During manufacturing of turbine or jet engines, combustion testing is required at different stages of development, manufacturing, installation, and deployment to ensure that the engine is working as designed and within desired tolerances. Multiple of such experiments are required for different operating points making this process very time consuming and expensive. Traditionally, test plans are created prior to actual experiments which consists of heuristics test blocks or groups of tests. These test blocks are generally created from expert judgment based on prior operational experience. However, these test plans may not be optimal as they are heuristically designed without rigorous statistical analysis of legacy and existing data. The obvious results is that traditional heuristic test plans may lead to inefficient and redundant allocation of resources.

Therefore, GE-IDACE can be employed to improve the test schedule. This happens by adaptively learning the system characteristics and performance with the underlying advanced surrogate model GEBHM as the function of input conditions using real-time data or even leverage historical experiments while also incorporating expert judgment. The test plan is hereby dynamically learned, compiled, ranked, and updated.

Ideally, historical data is available. The first step is to build GEBHM on this dataset. The input variables in this application are gas splits, loads, speed, firing temperature, etc., and the outputs are NOx emission, combustor instability, system dynamics, etc. The process followed with GE-IDACE is as shown in **Figure 2**. The steps are repeated as more data is added until necessary goals in the form of certification approval, optimum operating conditions, and/or constraints in the shape of time and budget are all met as an example.

The user typically possess desirability ratings for experimental outcomes. This desirability may include a factor, threshold, constraint, goal, or objective that is important to the user for testing such as emissions thresholds, maximum and minimum loads, efficiency ratings, among others. For example, the user would like to know the operating conditions at which the load is maximum while NOx exhaust and temperature are within some limits. The desirability is provided by the user as target values, target ranges, or by a custom function over the quantity of interest; please see Section 2.2.3 for a review.

With this introduction, in the following we demonstrate the impact of using GE-IDACE on combustion testing. A design space of four operating conditions x_1, x_2, x_3, and x_4 are explored, such that two performance parameters y_1 and y_2 stay within some thresholds defined as: $y_1 \in \left[y_1^{low}, y_1^{high}\right]$ and $y_2 \in \left[y_2^{low}, y_2^{high}\right]$. The goal is to design a test plan to maximize the number of experiments within said thresholds.

First, for later comparison, the traditional approach with one-factor-at-a-time designs are shown in **Figure 5**. Grey points indicate experiments out-of-bounds from a threshold perspective. Blue points met the conditions, i.e., they are within the blue delineated region of objective space. Out of a total of 69 experiments performed, 10 (14.5%) satisfied the desirability of y_1, 27 (39.1%) satisfied the desirability of y_2, and 35 (50.7%) satisfied the desirability of both y_1 and y_2.

Then, as an aim to improve this process, GE-IDACE was used to carry out a dynamic test plan. After each experiment, GEBHM was updated on the new data and the next point was picked based on the desirability with regards to the output responses. The corresponding output performance of these experiments and desirable regions is shown in **Figure 5B** to be compared with **Figure 5A**. Out of a total of 69 experiment performed, 25 (36.2%) satisfied the desirability of y_1, 40 (57.9%) satisfied the desirability of y_2, and 47 (68.2%) satisfied the desirability of both y_1 and y_2. The impact is that GE-IDACE increases the number of points in the desirable region by 20% with the same number of tests. Given the high cost of running these experiments, this easily translates to hundreds of thousands of dollars saved annually.

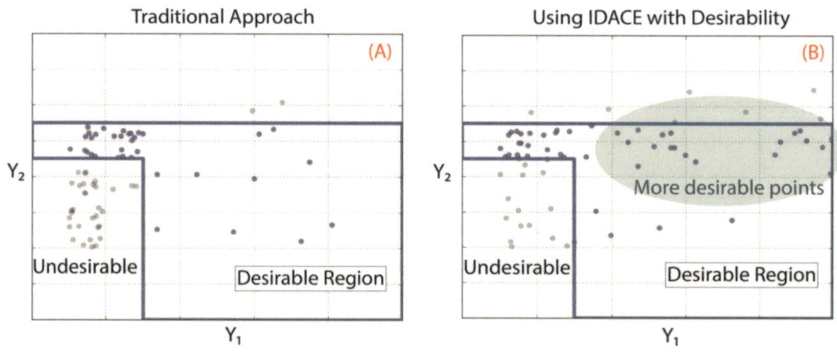

Figure 5.
(A) Results from the testing approach which does not utilize GE-IDACE. The plot shows the two-dimensional output space and the desirable region is delineated with a blue line and identifying text in the top right corner. Blue dots indicate experiments that met the desirability. Grey points did not meet desirabilities. (B) Results from the testing approach which utilizes GE-IDACE. The plot shows the two-dimensional output space and the desirable region is delineated with a blue line and identifying text in the top right corner. Blue dots indicate experiments that met the desirability. Grey points did not meet desirabilities. Comparing to (A), a higher fraction of points are blue and thus located in the desirable region.

In summary, the impact of GE-IDACE is clear to combustion testing. In fact, the application has now extending beyond just combustion testing: we are performing full-scale engine testing as well with this new strategy and it is being rolled out to multiple GE's businesses thus achieving severe cost reductions and better engineering designs.

3.3 Computational fluid dynamics for turbine design

Next, we consider the application of GE-IDACE to turbine design with CFD. So far in this chapter, we have covered real-world engineering applications. In the following, we demonstrate how GE-IDACE is positively impacting expensive computer simulations as well.

Aerodynamic optimization of a turbine involves dozens of variables, impacting everything from system level features through detailed airfoil properties. Two primary top-level considerations for the aerodynamic design of a turbine include vortexing and airfoil stack. Vortexing involves custom tailoring of the vane and rotor exit angle distributions. This establishes the radial distribution of work within the turbine stage. Vortexing affects local acceleration and mass flow distributions, and thus is a strong driver of secondary loss generation (endwall vortices). Airfoil stacking aerodynamically imposes body forces on the flow, further affecting the radial mass flow and work distributions. Stacking also strongly influences the generation of secondary loss. The general objective of a vortexing and stack optimization is to maximize turbine performance, usually through management of secondary loss growth, while also adhering to numerous constraints that ensure proper downstream performance and acceptable component life.

Before covering how GE-IDACE improved the optimization, consider the traditional approach as shown in **Figure 6**. The first stage vane is optimized using a component-specific space-filling DOE on which CFD is evaluated. These results are then used to build a surrogate model that characterizes a row-specific loss metric (relative total pressure loss or secondary kinetic energy, for example). A genetic algorithm (GA) is used to optimize a set of X's (defining a design point) describing the geometry for minimal loss (maximum efficiency) based on the surrogate model. This process is repeated for each subsequent row, with downstream components reacting to the results of the upstream row's optimized exit flow conditions.

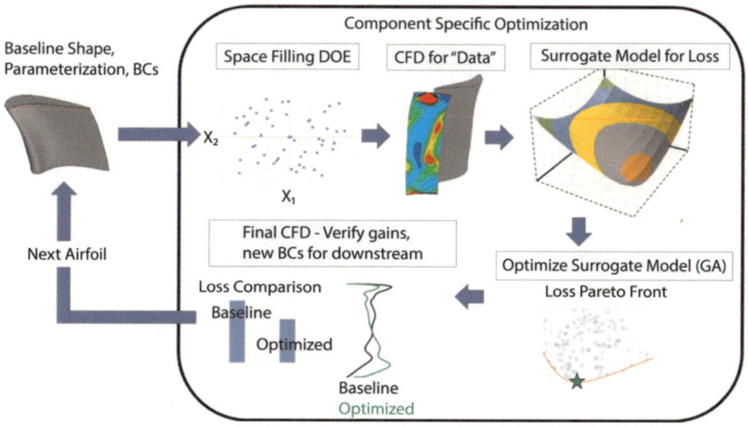

Figure 6.
Traditional approach to turbine blade optimization.

Compared to the traditional approach, GE-IDACE can help automate the geometry generation process to ensure efficient throughput. To accelerate build time from an X-specification to a CFD-ready geometry, a mesh morphing approach is implemented. Leveraging the block structured hexahedral mesh from the baseline geometry's CFD analysis, new cases re-stretch the inlet, exit, and passage blocks to produce a topologically identical mesh that conforms to the new 3D airfoil surface. The O block surrounding airfoil remains largely unchanged and translates with the new geometry. The baseline mesh is similar in fidelity to a typical "production" CFD analysis for turbine design. Surface y + is ~1 for all airfoil metal surfaces, and in total, the high pressure turbine (HPT) domain consists of ~9 million nodes. **Figure** 7 shows a representative example of the baseline grid and how it is morphed to an updated geometry.

All processes required to translate X's to CFD geometries are batch enabled, and each new CFD case requires ~15 min of wall clock time to generate. The CFD analysis is performed using GE's in-house CFD solver, TACOMA. TACOMA is a 2nd order accurate (in time and space), finite-volume, block-structured, compressible flow solver, implemented in Fortran 90. Stability is achieved via the JST scheme, and convergence is accelerated using pseudo-time marching and multi-grid techniques. The Reynolds Averaged Navier-Stokes (RANS) equations are closed via the k-ω turbulence model of Wilcox. Multi-row analysis is enabled through the use of mixing plane interfaces. Using 64 total CPU cores for the four-airfoil HPT domain, convergence is achieved in roughly 6 h.

The objective of the blade design task will be group efficiency. This metric is evaluated for each candidate point as a delta from a known baseline, which for this case is a modern two-stage Aviation HPT that already leverages results from prior optimization using the traditional techniques described earlier. All four HPT airfoils are considered in this optimization. To establish an entitlement performance, no constraints are imposed at this time to account for mechanical requirements or downstream component performance. Traditional space-filling DOEs for high dimensional problems require a large number of data points, and for a CFD-based study, an out-of-budget amount of computational resources. To manage these requirements, and to maintain design-cycle-relevant optimization times, advanced machine learning techniques are employed to intelligently guide the optimization process.

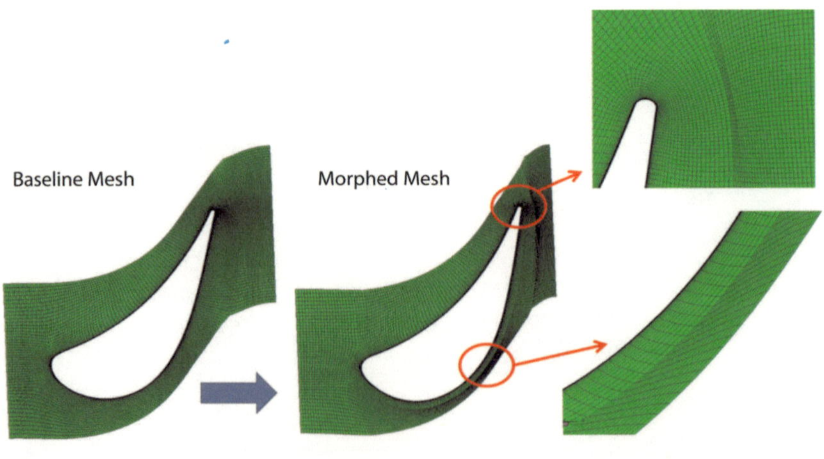

Figure 7.
Baseline and representative morphed mesh for an HPT vane.

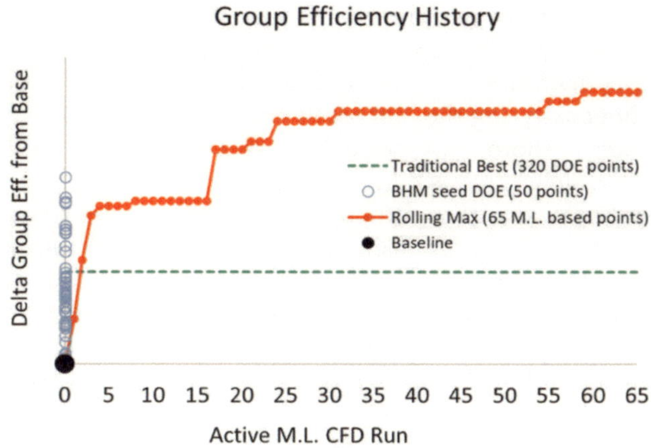

Figure 8.
Results of using GE-IDACE for turbine fan blade optimization. The traditional best shown with a green dashed line identifies the optimal design previously obtained using a mix of strategic designs and expert insights. GE-IDACE is the red full line and automates the design process and is clearly seen to outperform status quo. The initial design from which GEBHM is built from is shown as blue circles.

As a benchmark for the new process a 320-point space-filling OLH DOE was first created to cover all 32 HPT variables. Consistent with current practice, a radial basis function (RBF) surrogate model was fit to this data set, and GA optimization was performed on the RBF model. Modest gains over baseline were achieved from this approach.

For GE-IDACE, a more sparsely populated OLH DOE of 50 points was generated to seed the optimization. Leveraging GEBHM capabilities, additional DOE points were added only in areas of high error where the GEBHM model predicted high likelihood of progressing toward the objective—maximum delta group efficiency over the baseline. Through several rounds of intelligent incremental point addition, where each round included a refinement to the GEBHM fit, a new final optimal was established that exceeded the previous delta by roughly three times. Additionally, as shown in **Figure 8**, this much more favorable outcome was achieved with roughly a third of the computational resources.

4. Summary and future work

It has been demonstrated how advanced engineering tools centered around adaptive sampling in multi-objective space help achieve better engineering designs at highly-reduced cost. The underlying technologies are GEBHM and GE-IDACE which were covered first from a theoretical perspective. Then, applications in the areas of additive manufacturing, combustion testing, and computational fluid dynamics were considered. The impact of using GEBHM/GE-IDACE was clear and far surpassed status quo. At GE we consistently find a 30–90% resource cost reduction.

Before discussing future work, we first cover some of the main limitations of the GE-IDACE tool. Fundamentally, GE-IDACE treats the computer experiment as a black box function, i.e., it only sees inputs to the code and the corresponding outputs. In some cases, this information is all we are able to leverage, but in other situations we may have additional insights which, if taken advantage of, could speed up the optimization. For example, gradient information could be available from the experiments too. Furthermore, the GE-IDACE approach is "greedy," i.e., it selects the next input point from the design space which is predicted to give the best immediate outcome with the current state of knowledge. This approach might not be the optimum strategy in the long-term. Worded differently, under a budget, there could exist a possibility that one can reach a better overall solution with fewer experiments without selecting the highest EI point in each intermediate step. Finally, it is difficult to thoroughly parallelize experiments with GE-IDACE as it is a sequential process which requires data acquisition and model-updating as the experimental results are available, although some approximate schemes exist [50].

In terms of future work and further improvements, we demonstrate in Ref. [51] that Particle Swarm Optimization performs very well for EI computation. A lot of exciting opportunities exist for GEBHM and GE-IDACE to further improve the engineering design process and remain to be discovered. In recent work, we demonstrate how to use GE-IDACE with multi-fidelity data sources (simulation vs. experiments, e.g.,) [52] and how to leverage legacy data from other designs into the GEBHM modeling process [11], to reduce the cost of running tests for new engine designs. In terms of future work, GEBHM can be extended to operate fluently across any type of data in terms of dimensionality and number of points. This way, all the benefits of GEBHM and GE-IDACE can be leveraged at any scale. Toward this, an initial exploration of a parallelizable way to fit the GEBHM is found in Ref. [53]. This extends the size of datasets which GEBHM can fit by a factor of 5–10.

Acknowledgements

The authors acknowledge support from GE Research and are grateful to Tom Vandeputte at GE Research for details on the CFD modeling.

Author details

Jesper Kristensen, Waad Subber, Yiming Zhang, Sayan Ghosh*, Natarajan Chennimalai Kumar, Genghis Khan and Liping Wang
Probabilisitcs Design and Optimization, GE Research, Niskayuna, New York, USA

*Address all correspondence to: sayan.ghosh1@ge.com

References

[1] Rasmussen CE. Gaussian Processes in Machine Learning. 2004. pp. 63-71

[2] Bishop CM. Pattern Recognition and Machine Learning. Springer; 2006

[3] Gelman A, Carlin JB, Stern HS, Dunson DB, Vehtari A, Rubin DB. Bayesian Data Analysis. Chapman and Hall/CRC; 2013

[4] Murphy KP. Machine Learning: A Probabilistic Perspective. MIT Press; 2012

[5] Seeger M. Gaussian processes for machine learning. International Journal of Neural Systems. 2004;**14**(02):69-106

[6] Kennedy MC, O'Hagan A. Bayesian calibration of computer models. Journal of the Royal Statistical Society, Series B: Statistical Methodology. 2001;**63**(3): 425-464

[7] Kristensen J, Asher I, Ling Y, Ryan K, Subramaniyan A, Wang L. Predictive analytics with an advanced Bayesian modeling framework. MODSIM World. 2017

[8] Marzouk YM, Najm HN, Rahn LA. Stochastic spectral methods for efficient Bayesian solution of inverse problems. Journal of Computational Physics. 2007; **224**(2):560-586

[9] Subber W, Salvadori A, Lee S, Matous K. Uncertainty Quantification of the Reverse Taylor Impact Test and Localized Asynchronous Space-Time Algorithm. Bulletin of the American Physical Society; 2017. p. 62

[10] Sacks J, Welch WJ, Mitchell TJ, Wynn HP. Design and analysis of computer experiments. Statistical Science. 1989;**4**(4):409-423

[11] Ghosh S, Asher I, Kristensen J, Ling Y, Ryan K, Wang L. Bayesian multi-source modeling with legacy data. In: 2018 AIAA Non-Deterministic Approaches Conference. 2018. p. 1663

[12] Bilionis I, Zabaras N, Konomi BA, Lin G. Multi-output separable Gaussian process: Towards an efficient, fully Bayesian paradigm for uncertainty quantification. Journal of Computational Physics. 2013;**241**: 212-239

[13] Simpson T, Toropov V, Balabanov V, Viana F. Design and analysis of computer experiments in multidisciplinary design optimization: A review of how far we have come-or not. In: 12th AIAA/ISSMO Multidisciplinary Analysis and Optimization Conference. 2008. p. 5802

[14] Viana FAC, Venter G, Balabanov V. An algorithm for fast optimal Latin hypercube design of experiments. International Journal for Numerical Methods in Engineering. 2010;**82**(2): 135-156

[15] Kristensen J, Ling Y, Asher I, Wang L. Expected-improvement-based methods for adaptive sampling in multi-objective optimization problems. In: ASME. International Design Engineering Technical Conferences and Computers and Information in Engineering Conference, Volume 2B: 42nd Design Automation Conference, Design Automation Conference; Charlotte, North Carolina, USA; August 21–24, 2016

[16] Chennimalai Kumar N, Subramaniyan AK, Wang L. Improving high-dimensional physics models through Bayesian calibration with uncertain data. In: ASME Turbo Expo 2012: Turbine Technical Conference and Exposition. American Society of Mechanical Engineers; 2012. pp. 407-416

[17] Kristensen J, Asher I, Wang L. Polynomial representation of the gaussian process. In: ASME. International Design Engineering Technical Conferences and Computers and Information in Engineering Conference, Volume 1B: 38th Computers and Information in Engineering Conference; Quebec City, Quebec, Canada; August 26–29, 2018

[18] Ling Y, Ghosh S, Asher I, Kristensen J, Ryan K, Wang L. An intelligent sampling framework for multi-objective optimization in high dimensional design space. In: 2018 AIAA/ASCE/AHS/ASC Structures, Structural Dynamics, and Materials Conference, AIAA SciTech Forum, American Institute of Aeronautics and Astronautics; Kissimmee, Florida; 8–12 January, 2018

[19] Ling Y, Ryan K, Asher I, Kristensen J, Ghosh S, Wang L. Efficient robust design optimization using gaussian process and intelligent sampling. In: 2018 Multidisciplinary Analysis and Optimization Conference, 2018 Multidisciplinary Analysis and Optimization Conference, AIAA AVIATION Forum. Atlanta, Georgia: American Institute of Aeronautics and Astronautics; 2018

[20] Ryan KM, Kristensen J, Ling Y, Ghosh S, Asher I, Wang L. A Gaussian process modeling approach for fast robust design with uncertain inputs. In: ASME. Turbo Expo: Power for Land, Sea, and Air, Volume 7A: Structures and Dynamics; Oslo, Norway; June 11–15, 2018

[21] Srivastava A, Subramaniyan AK, Wang L. Analytical global sensitivity analysis with Gaussian processes. AI EDAM. 2017;**31**(3):235-250

[22] Srivastava A, Subramaniyan AK, Wang L. Hybrid bayesian solution to NASA langley research center multidisciplinary uncertainty quantification challenge. Journal of Aerospace Information Systems. 2015; **12**(1):114-139

[23] Srivastava A, Subramaniyan AK, Wang L. Variance based global sensitivity analysis for uncorrelated and correlated inputs with gaussian processes. In: ASME Turbo Expo 2015: Turbine Technical Conference and Exposition. American Society of Mechanical Engineers; 2015. pp. V07AT29A004-V07AT29A004

[24] Subramaniyan AK, Kumar NC, Wang L. Probabilistic validation of complex engineering simulations with sparse data. In: ASME Turbo Expo 2014: Turbine Technical Conference and Exposition. American Society of Mechanical Engineers; 2014

[25] Wang L, Fang X, Subramaniyan A, Jothiprasad G, Gardner M, Kale A, et al. Challenges in uncertainty, calibration, validation and predictability of engineering analysis models. In: ASME 2011 Turbo Expo: Turbine Technical Conference and Exposition. American Society of Mechanical Engineers; 2011. pp. 747-758

[26] Zhang Y, Ghosh S, Asher I, Ling Y, Wang L. Learning uncertainty using clustering and local gaussian process regression. In: AIAA Scitech 2019 Forum. 2019. p. 1730

[27] Chib S, Greenberg E. Understanding the Metropolis-Hastings algorithm. The American Statistician. 1995;**49**(4):327-335

[28] Hastings W. Monte Carlo sampling methods using Markov chains and their applications. Biometrika. 1970;**57**(1): 97-109

[29] Mengersen KL, Robert CP, Guihenneuc-Jouyaux C. MCMC convergence diagnostics: A review. Bayesian Statistics. 1999;**6**:415-440

[30] Jones DR, Schonlau M, Welch WJ. Efficient global optimization of

expensive black-box functions. Journal of Global Optimization. 1998;**13**(4):455-492

[31] Couckuyt I, Deschrijver D, Dhaene T. Fast calculation of multiobjective probability of improvement and expected improvement criteria for pareto optimization. Journal of Global Optimization. 2014;**60**(3):575-594

[32] Emmerich M, Klinkenberg JW. The computation of the expected improvement in dominated hypervolume of Pareto front approximations. In: Rapport Technique. Vol. 34. Leiden University; 2008. pp. 7-13

[33] Emmerich MT, Deutz AH, Klinkenberg JW. Hypervolume-based expected improvement: Monotonicity properties and exact computation. In: 2011 IEEE Congress of Evolutionary Computation (CEC). IEEE; 2011. pp. 2147-2154

[34] Keane AJ. Statistical improvement criteria for use in multiobjective design optimization. AIAA Journal. 2006; **44**(4):879-891

[35] Pandita P, Bilionis I, Panchal J. Extending expected improvement for high-dimensional stochastic optimization of expensive black-box functions. Journal of Mechanical Design. 2016;**138**(11):111412

[36] Pandita P, Bilionis I, Panchal J. Deriving information acquisition criteria for sequentially inferring the expected value of a black-box function. In: arXiv preprint arXiv:1807.09979; 2018

[37] Pandita P, Bilionis I, Panchal J, Gautham BP, Joshi A, Zagade P. Stochastic multi-objective optimization on a budget: Application to multi-pass wire drawing with quantified uncertainties. In: arXiv preprint arXiv: 1706.01665; 2017

[38] Ponweiser W, Wagner T, Vincze M. Clustered multiple generalized expected improvement: A novel infill sampling criterion for surrogate models. In: 2008 IEEE Congress on Evolutionary Computation (IEEE World Congress on Computational Intelligence). IEEE; 2008. pp. 3515-3522

[39] Wagner T, Emmerich M, Deutz A, Ponweiser W. On expected-improvement criteria for model-based multi-objective optimization. In: Parallel Problem Solving from Nature, PPSN XI. Berlin, Heidelberg: Springer Berlin Heidelberg; 2010. pp. 718-727

[40] Lotov AV, Miettinen K. Visualizing the Pareto frontier. In: Multiobjective Optimization. Heidelberg: Springer, Berlin; 2008. pp. 213-243

[41] Wilson B, Cappelleri D, Simpson TW, Frecker M. Efficient Pareto frontier exploration using surrogate approximations. Optimization and Engineering. 2001;**2**(1):31-50

[42] Frazier WE. Metal additive manufacturing: A review. Journal of Materials Engineering and Performance. 2014;**23**(6):1917-1928

[43] Forrester A, Sobester A, Keane A. Engineering Design Via Surrogate Modelling: A Practical Guide. John Wiley & Sons; 2008

[44] Emmerich M, Beume N, Naujoks B. An EMO algorithm using the hypervolume measure as selection criterion. In: International Conference on Evolutionary Multi-Criterion Optimization. Berlin, Heidelberg: Springer; 2005. pp. 62-76

[45] Bader J, Zitzler E. HypE: An algorithm for fast hypervolume-based many-objective optimization. Evolutionary Computation. 2011;**19**(1): 45-76

[46] Messac A, Ismail-Yahaya A. Multi objective robust design using physical

programming. Structural and Multidisciplinary Optimization. 2002; **23**(5):357-371

[47] Zitzler E, Brockhoff D, Thiele L. The hypervolume indicator revisited: On the design of Pareto-compliant indicators via weighted integration. In: International Conference on Evolutionary Multi-Criterion Optimization. Berlin, Heidelberg: Springer; 2007. pp. 862-876

[48] Aggour KS, Gupta VK, Ruscitto D, Ajdelsztajn L, Bian X, Brosnan KH. Artificial intelligence/machine learning in manufacturing and inspection: A GE perspective. MRS Bulletin. 2019;**44**(7): 545-558

[49] Wang L, Asher IM, Ling Y, Srivastava A, Subramaniyan AK. General Electric Co, 2018. Systems and method for dynamic combustion tests. U.S. Patent Application 15/445,334

[50] Wang J, Clark SC, Liu E, Frazier PI. Parallel bayesian global optimization of expensive functions. arXiv preprint arXiv:1602.05149. 2016

[51] Zhang Y, Kristensen J, Ghosh S, Vandeputte T, Tallman J, Wang L. Finding maximum expected improvement for high-dimensional design optimization. In: AIAA Aviation 2019 Forum. 2019. p. 2985

[52] Ghosh S, Kristensen J, Zhang Y, Subber W, Wang L. A Strategy for Adaptive Sampling of Multi-fidelity Gaussian Process to Reduce Predictive Uncertainty. [online] arXiv.org. 2019. Available from: https://arxiv.org/abs/1907.11739

[53] Pandita P, Kristensen J, Wang L. Towards Scalable Gaussian Process Modeling. [online] arXiv.org. 2019. Available from: https://arxiv.org/abs/1907.11313 [Accessed: 29 July 2019]

Chapter 3

Design for Sustainability with Biodegradable Composites

Dina Fouad and Mahmoud Farag

Abstract

Many of the petroleum-based materials and products are causing problems with sustainability of resources and disposal at the end of their lives. Such problems can be solved if biodegradable materials from renewable resources are used in product design. For a material to be fully biodegradable, all its constituents must be biodegradable and should come from renewable resources if it is to be sustainable. Starch-plant fiber composites satisfy both conditions. In addition to their environmental benefits, materials from renewable resources can also be economically advantageous in certain applications, such as motorcar and packaging industries. This chapter starts with a review of the characteristics of biodegradable materials and uses case studies to illustrate their use in the design of sustainable products. The concept of design for a life (DFL), in which the material used in making a given product that will biodegrade at the end of its useful life, will also be explored.

Keywords: sustainability, design for a life (DFL), biodegradable composites, natural polymers, natural fibers, degradation, economics of sustainable designs

1. Introduction

Recently, several reviews have been published to report on the advancement in the fabrication and superior properties achieved in biodegradable materials [1–5]. Such growing global notion and urgency toward the need for biodegradable sustainable alternatives to petroleum-based synthetic plastics have stemmed from the increased environmental awareness, depletion of the scarce nonrenewable resources, as well as the implementation of stringent governmental regulations in several countries [5–10]. They mostly emphasized on the suitability of bio-based polymer composites as substitutes for conventional synthetic polymers. The main problem stems from the fact that plastics have become one of the major pollutants of current times. Due to the ubiquity of nondegradable plastics associated with products used in everyday life and the deficiency in creating a proper recycling infrastructure, plastic waste has proliferated over the years and accumulated in landfills and oceans causing severe ecological and environmental problems across the globe [10]. Another concern is the possibility of such waste releasing toxins in landfills and reaching food resources, which has a negative impact on human health [11, 12]. Consequently, a high demand for biodegradable alternatives has arisen in different fields and industries as an attempt to achieve a more environmentally friendly approach in product manufacturing and design [7]. Accordingly, the European bioplastics report published in 2018 has forecasted more than a 20% growth in the production of bioplastics by the year 2023 [13]. Concurrently, the demand and production of plastics in general continue to rise, where it is expected to reach staggering 540 million tons in 2020, where only 2.2 million tons are estimated to be of bio-based resources [13, 14]. Despite the promising growth rates in the field of bioplastics, comparison demonstrates that the rate of growth is still very small in contrast to that anticipated for traditional plastics. The transition and adaptation of bioplastics is generally impeded by their higher cost of production and lower durability as opposed to their conventional counterparts [8, 9, 15]. Therefore, it is important for the degradable plastic alternatives to be designed so as to offer the same functionality as the original synthetic plastics for the required service life and at a competitive cost [8]. Here, the cost is only justified when the cost of sustainability is taken into consideration and not only that of production [15]. On the other hand, in spite of the challenges faced, several industries such as the packaging, automation, consumer goods, and biomedical fields have shown encouraging implementations of biodegradable alternatives in their plastic-based commodities, taking the necessary preliminary steps towards the commercialization of bio-based and biodegradable plastics [7, 13, 16]. Products that are produced using such materials are designed to biodegrade at the end of their useful life using the design for a life (DFL) approach which is a crucial element in the successful transition to sustainable bioplastics [17]. For instance, a packaging material can start to biodegrade soon after the consumption of its contents with a supplementary benefit of avoiding any harmful contaminants that could have leached to the food content from the synthetic counterparts, which indicates a clear environmental advantage. Ecological and sustainable advantages are also achieved when they are applied for components of automobiles to achieve an eco-friendly design [18]. Thus, in order to achieve a successful implementation of the DFL approach, a thorough understanding of the different biodegradable polymers and composites available is needed along with their properties, methods of production, and degradation to properly evaluate and asses its life cycle and positive impact. This chapter presents an overview on the different types of biodegradable polymers and composites highlighting their main characteristics and advantages. Moreover, the use of the design for a life approach will be elucidated using a case study in the field of automation.

2. Biodegradable polymers: classifications and properties

Biodegradable plastics can be derived from either synthetic or natural resources and are commonly referred to as "biopolymers" [5]. They are defined as polymers

Category	Type	Source	Source type	Method of production	Example
Synthetic	Bio-polyesters	Bio-derived	Renewable	Chemical polymerization of bio-monomers	PLA
		Synthetic monomer	Nonrenewable	Polymerization of lactide	PLA
		Bio-chemosynthetic monomer	Renewable	Biosynthesis of polymers in microorganisms	PHA
		Synthetic monomer	Nonrenewable	Enzymatic catalyzed polymerization	PCL
Natural	Agropolymer	Biomass products	Renewable	Fragmentation of biomass	Starch

Table 1.
Classification of biodegradable polymers [7, 19–22].

that naturally degrade and assimilate in the environment into water (H_2O) and carbon dioxide (CO_2) by means of microorganisms [23]. In regard to the bio-polyesters, their hydrolysable ester bonds are what make them biodegradable, while for the natural polymers, the process is usually through hydrolysis [7]. Moreover, the means of fabrication are categorized into three main classes: (1) chemical polymerization of monomers originating from biological processes such as in the case of polylactic acid (PLA), (2) chemosynthesis of the polymers in microorganisms such as polyhydroxyalkanoate (PHA), and (3) modification of natural polymers, i.e., starch [5]. **Table 1** summarizes the different typologies used to categorize the different types along with their source and methods of production. Also, characteristic examples of each type are indicated, and their main features highlighted.

2.1 Bio-polyesters: synthetic polymers

Among the most representative of the synthetic polymers are the aliphatic bio-polyesters (listed in **Table 1**), PLA, PHA, and polycaprolactone (PCL) [18]. Poly-condensation of bifunctional monomers and ring-opening polymerization processes are commonly used to yield high molecular weight polymers [7]. A comprehensive overview of the different chemical synthesis methods used to synthesize them is reviewed in [23]. The prime interest in this class of materials is due to the fact that they exhibit mechanical properties equivalent to petroleum-based polymers such polyethylene (PE) and polypropylene (PP) [19]. A summary of the reported mechanical and physical properties of the bio-polyesters and natural biopolymers in comparison to the polyolefin low-density polyethylene (LDPE) is provided in **Table 2**.

2.1.1 Polylactic acid

Polylactic acid is a high molecular weight, crystalline thermoplastic obtained from the ring polymerization of lactide [20]. It was first synthesized in 1931 by a DuPont scientist and was derived from agricultural products such as corn [22]. The typical glass transition temperature (T_g) falls in between 40 and 70°C, while the melting temperature (T_m) is between 130 and 180°C, as referred in **Table 2**. Additionally, it exhibits high strength, where the average tensile strength is 50 MPa compared to 14 MPa of polyolefins such as LDPE.

Table 3 demonstrates the essential differences between the different biopolymers with regard to their cost, mechanical properties, hydrophilicity, and biodegradation rate. PLA is a hydrophobic polymer due to the methane side group present along the chain's backbone. Thus, it is more resistant to hydrolysis than PHAs, and hence, their biodegradation rate is relatively slow [7]. Moreover, the hydrolytic degradation process needs to be catalyzed at high temperatures, normally in the

Polymer	T_g (°C)	T_m (°C)	UTS (MPa)	ε (%)	Degradation time (months)	Reference
LDPE	−100	98–115	8–20	100–1000	NA	[20]
PCL	−60	58–63	4–28	700–1000	>24	[20, 24]
PLA	40–70	130–180	48–53	5–8	12–16	[20, 24]
PHA	−30–10	70–170	18–24	3–25	1–2	[20, 26]
Starch	60–80	—	2.6	47	Bulk	[6, 25]

Table 2.
Physical and mechanical properties of biodegradable polymers compared to non-biodegradable LDPE.

range of 60°C. This means that it is not compostable at home, for instance, and requires a specific compostable environment [22]. Despite its high strength, PLA is limited due to its brittleness thermal instability [20]. Additionally, PLA polymers are derived from nonrenewable resources that make their sustainability a questionable matter. While other semisynthetic variants are fabricated; however, their use would not be favorable due to being partially degradable [7].

2.1.2 Polyhydroxyalkanoate (PHA)

PHA is a microbial polymer which retains close properties to non-biodegradable thermoplastics such as LDPE, as shown in **Table 2** [27]. Its discovery and synthesis began during the early twentieth century [22]. It possesses desirable physical and mechanical characteristics such as high melting temperature; the tensile strength is slightly higher than that of LDPE with an average of 21 MPa. Unlike the aforementioned PLAs, PHAs exhibit good impermeability to water as well as rapid biodegradation properties, as shown in **Table 3**. They are a class of biopolymers which biodegrade by microorganisms. Additionally, they are considered thermoplastic and could be easily processed using existing molding and extrusion technologies known for petrochemical-based polymers [20]. Their rapid degradation under various environments is considered its main attribute in contrast with other alternatives, not to mention that they are considered as a sustainable substitute for being biocompatible and biorenewable as it originates from plant oils and sugars [27]. However, methods of synthesis are highly costly, and until a cost-effective method is derived, PHA commercialization would be limited and economically unjustified.

2.1.3 Polycaprolactone

PCL is a semicrystalline polymer fabricated by means of ring polymerization of caprolactone in the presence of a catalyst [1, 7]. It is considered a synthetic biopolymer fabricated from a nonrenewable resource, and similar to PLA, PCL is a hydrophobic polymer with a low degradation rate [18]. Additionally, compared to PCL, the average tensile strength is lower, but it is important to note its high elongation at break being comparable to that of LDPE. Nevertheless, its high cost of processing and non-renewability are major drawbacks in the context of sustainable design.

To that end, in spite of the several advantages offered by synthetic polymers such as compatible properties and easy processing, they are considered very expensive to produce [6]. Also, from a sustainability point of view, they offer a weak competition, with some being nonrenewable such as PCL and partially degradable as PLA. This dictates the importance of the agro-/natural polymers that are inherently both biodegradable and biorenewable as well as being cheaper to

Polymer	Cost of processing	Mechanical properties	Impermeability to water	Degradation rate
PCL	High	Moderate	Good	Slow
PLA	High	High and brittle	Good	Slow
PHA	High	Moderate	Good	Rapid
Starch	Low	Poor	Moderate	Fast

Table 3.
Comparison between biodegradable polymers [16, 18, 20, 21].

produce. Owing to their attractive attributes, this category provides a promising sustainable candidate for green product design over their synthetic equivalents.

2.2 Agropolymers: natural polymers

Natural polymers form during the ecological growth cycle of living organisms [7]. They are mainly derived from biomass fragmentation processes, where polysaccharides are classified as the most representative family of natural polymers.

2.2.1 Polysaccharides: starch polymers

The main polysaccharides explored in various applications are starch and cellulose-based derivatives [7]. Owing to being abundant, low in cost, and biodegradable, starch-based polymers are among the most extensively studied biodegradable polymer and are considered one of the most favorable candidates for sustainable materials [2, 6, 7, 14–16, 19, 20, 25, 28, 29]. Starches are hydrophilic carbohydrate materials that are regenerated by photosynthesis from plants such as wheat, corn, rice, and potato [6, 28].

Starch is primarily composed of two glucose homopolymers: (1) linear amylase and (2) highly branched amylopectin [1, 16]. Different sources yield different proportions of the homopolymers in the range of 10–25% amylose and 75–90% amylopectin [6]. This leads to variable properties, where high amylose content in starch leads to an improvement in mechanical properties such strength and elongation [7, 20]. Additionally, the hydroxyl side groups present in the polymeric chain aid in the rapid biodegradation of the biopolymer [6]. The polymer is considered as highly sustainable, where it is worth noting that during the natural assimilation process, starch is hydrolyzed into glucose that is further metabolized into CO_2 and H_2O. Afterwards, an ecological equilibrium is created, whereas aforementioned the starch is regenerated by the natural photosynthesis process of plants as they absorb the processed CO_2 [6, 27]. Nevertheless, it is important to note that native starch by itself cannot be processed and it must undergo a modification process to improve its processability.

2.2.2 Modification of natural starch: gelatinization to form thermoplastic starch (TPS)

The modification process initiates by applying thermomechanical processing to the starch granules mixed with water at temperatures in the range of 90–180°C, which causes expansion and disruption of the granules as a means of transforming the semicrystalline structure into an amorphous thermoplastic starch [28]. This process is referred to as "gelatinization," and at this stage, the starch is difficult to process, and the addition of a plasticizer such as glycerol or other polyhydroxy compounds is needed to reduce the glass transition temperature (T_g) and improve its flow properties and mechanical properties as reported by Elsayed et al. [28], Ibrahim et al. [29], and Mehanny et al. [30], where optimum conditions were found at glycerin content of 30 wt.%. Additionally, Vroman and Tighzert reported—in their review on biodegradable polymer—an improvement in the flexibility and elongation properties at glycerol contents higher than 20% [7]. Thus, the resultant properties and loss of crystallinity is a function of the type of starch used and supplied water and heat during the gelatinization process [16, 28]. Concurrently, the processing technique plays a vital role in the crystallinity and mechanical properties of TPS, where the shear stresses applied during the extrusion process allows for an efficient transfer of the water into the molecules, and also the use of injection molding leads to a more amorphous structure and ductile properties [16].

Additionally, it is argued that the addition of plasticizers retards the retrograding process that takes place as the polymer recrystallizes and becomes brittle with time [31]. However, the pure thermoplastic starch still acquires properties similar to that of native starch such as poor mechanical properties and high hydrophilicity [2, 20]. This is due to the fact that the plasticizers themselves increase the hydrophilic nature of the polymer and results in higher water permeability [14, 31]. This leads to thermal instability and the loss of mechanical properties [2].

2.3 Modification of thermoplastic starch

2.3.1 Acetylation

To improve the properties of TPS, several methods were devised. The first technique is known as acetylation, where starch acetate is fabricated through the chemical mixture with pyridine and acetic acid [7]. The resultant polymer has a high content of the linear amylose polymer that is less hydrophilic, thus overcoming the permeability issue exhibited by pure TPS [7].

2.3.2 Grafting

Another powerful technique devised is grafting or copolymerization. Examples include grafting synthetic bio-polyesters such as PCL and PLA to the starch by a chemical bond [6]. However, it is argued that the rate of biodegradability is sacrificed under these conditions as the chains *will not* assimilate in nature readily nor easily [7].

2.3.3 Blending

At first, scientists used to blend starch with polyolefin synthetic polymers to achieve desirable superior properties; however these systems are partially biodegradable and thus are regarded unacceptable from a sustainability point of view [32]. Thus, the use of only biodegradable synthetic polymers is restricted while using this technique. A thorough review is provided in [7]. The most common components to blend with starch are the aliphatic bio-polyesters such as PLA, PCL, and PHA [33]. The resultant material achieves improved properties and cost competitiveness. On the other hand, a major shortcoming is reported, where it is outlined that starch and many polymers are immiscible, which, thereafter, causes these blends to become weak and eventually deteriorate [6].

Thereupon, it could be inferred that a large range of properties could be tailored among these polymers for specific applications. However, each of them exhibits a variety of limitations which restricts their use and applicability across many fields, and from a sustainability viewpoint, starch-based polymers still provide the best alternative especially if their shortcomings are overcome. Their main advantageous features rely on being the lowest cost material compared to other biodegradable polymers, which are processed by existent processing techniques used for conventional polymers. Also they are both renewable and biodegradable,where several studies have reported an immense improvement in the mechanical and physical properties of modified TPS. Different techniques such as blending, grafting, and acetylation have been implemented to improve properties; however they affect the biodegradability of TPS which is its key successor that establishes it as the front runner in the race toward achieving sustainable alternative materials [7]. Subsequently, it has been reported in the literature that tailoring starch-based composites that are dependent on natural resources yields optimum results while still preserving the biodegradability nature of the polymer [6]. Given their favorable potential, the nature of these composites will be discussed in Section 3 along with introducing the different types and characteristics.

3. Biodegradable starch-based composites

A biocomposite polymer is classified as a material which combines a biodegradable polymer as its matrix and a biodegradable filler as the reinforcement [18]. Such composites are also commonly known as "green composites," and as the focus in this chapter is sustainability, the use of natural fibers as fillers will be the only class investigated [34]. Several researchers have demonstrated a high compatibility between starch- and natural-based fibers such as cellulose derivatives [2, 14, 16, 18, 28, 29, 33]. Considerable improvement in the mechanical properties of starch-based composites coupled with a reduction in water permeability has been reported. Additionally, Reddy et al. has reported the use of nano-fillers specifically cellulose-based in the fabrication of green composites, where significant enhancement in properties is anticipated [35]. The classification of natural fibers is presented in Section 3.1, and an overview of their impact as reinforcements on the TPS matrix is provided in Section 3.2.

3.1 Natural fibers

Fiber fillers are added as the source of reinforcement and load bearing component within the composite matrix. They are of either natural or synthetic origin such as plants and carbon, respectively. However, natural fibers offer several benefits over their synthetic counterparts, one of them being that they are essentially biodegradable which is considered as a merit for the environment [8]. Additionally, high specific properties such as strength and low density along with being renewable and low in cost have led to their emergence as excellent substitutes for the man-made competitors [4, 36]. The fiber strength comes from the strong inter- and intramolecular bonds that make the fiber stiff and rigid developing intertwined threadlike structures [31]. In addition to the strong bonds, the higher the crystallinity of the filler material, the less exposed areas of the matrix that would absorb water and moisture. Each type differs slightly in their characteristics; there are three major classes upon which this family of fibers are classified: (1) plant-based fibers, usually referred to as bast fibers and are extracted from the outer bark of plant stems, such as flax, jute, and hemp; (2) leaf fibers, which are hard and strong fibers obtained from leaf tissues such as in the case of sisal and pineapple; and finally (3) seed fibers such as cotton and coir [4, 16]. Other types are extracted from wood or grass [4]. **Table 4** illustrates the mechanical properties of the characteristic natural fibers commonly used for each category compared to carbon fibers. It can be observed that the plant-based flax fibers exhibit the highest strength with a maximum of 1500 MPa, while that for the remaining bast and leaf fibers is less than 1000 MPa. Nevertheless, they all show high specific strength and specific stiffness properties compared to carbon fibers, where the specific gravity of carbon is much higher than the natural counterparts.

Furthermore, it is important to note that natural fibers have wax on its surfaces and other elements such as lignin and hemicellulose, which leads to difficulty in the adhesion of the matrix to the fibers. Therefore, to improve the poor linkage and adhesion problem, the fibers undergo a surface chemical treatment before synthesis with the matrix, which also aim to reduce the fiber permeability to water [16]. Most treatments work on removing the hydrogen bonds on the surface so as to make it hydrophobic and to improve the surface roughness.

Category	Type	UTS (MPa)	ε (%)	E (GPa)	Specific gravity (g/cc)
Bast fibers	Hemp	270–900	1.6	24–90	1.4–1.5
	Flax	345–1500	2.7–3.2	27–100	1.4–1.5
	Jute	393–800	1.16–1.5	13–55	1.3–1.49
Leaf fibers	Sisal	468–700	3–7	9.4–22	1.3
Seed fibers	Coir	131–220	15–40	4–6	1.15–1.46
Synthetic	Carbon	2500	1.4–1.8	425	1.9

Table 4.
Mechanical properties of natural fibers compared to carbon fibers [16, 34].

3.2 Natural fiber-reinforced starch-based composites: performance evaluation

The main goal driving the fabrication of the natural fiber-reinforced starch-based composites is overcoming the limitations of TPS and attaining better mechanical and physical properties while still retaining the biodegradability attribute of natural materials. This section attempts to evaluate the performance of the different variations of the green composite based on the improvements achieved in the mechanical properties, thermal stability, and biodegradation rates compared to that of the starch matrix. **Table 5** summarizes the outcome of the studies performed on different combinations of natural fiber and starch. Key findings and conclusions are drawn from comparing the data tabulated and will be discussed in the following sections, based on which the DFL approach will be highlighted.

3.2.1 Mechanical properties

As observed from **Table 5**, there is a general increasing trend in the tensile strength as the fraction of fiber increases in the composite compared to that of pure thermoplastic starch. This is demonstrated in the case of adding flax fibers to TPS, where the tensile strength increased from 50 to 60 MPa as a function of increasing the fiber content from 40 to 50%, respectively. However, increasing the content beyond certain percentages, the opposite occurs where the properties deteriorate instead of improving. This is slightly observed post increasing the flax fiber content to 80%, where the tensile strength reduced to 55 MPa. However, this phenomenon was clearly observed when using date palm fibers, where increasing the fiber composition from 50 to 80% led to a significant 60% decrease in the tensile strength from 32.7 to 12 MPa, respectively. Moreover, in spite of following similar trends, different fibers possess variable properties which eventually lead to major differences in the properties attained. It could be remarked that flax-based starch composites acquire the highest in tensile properties coupled with the highest ductility among other composites. It is clear that the strong flax fibers have imparted their high strength properties (listed in **Table 4**) to the starch matrix and produces a high strength composite with desirable properties. Additionally, the hybrid between date and flax fibers has led to an increase in the tensile strength equivalent to the average increase attained from each type—at the same composition—separately.

3.2.2 Thermal stability

Another upward trend is attained in the thermal stability which incrementally increases as a function of increasing the fiber content. Due to the organic nature of the biocomposite constituents, heat application is expected to cause changes in their physical and chemical properties [16]. Thus, thermal stability is tested through a

Fiber	Composition (%)	UTS (MPa)	ε (%)	E (GPa)	Weight loss (%)/week	Temp. at 10% weight loss (°C)	Reference
TPS	0	3.8	138	0.5	30	192	[28]
Flax	40	50	—	3.5	16	—	[28]
	50	60	5.7	4.3	15.4	229	[2]
	80	55	—	2	5	251	[28]
	50 (unidirectional)	131	5.8	7.5	15.4	—	[2]
Palm	50	28.2	1.82	3.85	18.6	—	[2]
Banana	50	25.4	2.03	3.71	20.3	—	[2]
Bagasse	50	29.8	3.27	3.23	20	—	[2]
Date	50	32.7	—	2.8	18	232	[29]
	80	12	—	7	10	250	[29]
Hybrid	25(Date) & 25 (flax)	43	—	—	—	—	[29]
Sisal	20	2.8	2	151	—	—	[38]
Hemp	20	4	3.4	182	—	—	[38]
Short fiber-cellulose	15	15.43	6.08	364.9	—	350	[39]
Jute	12.5	5.5	—	—	—	—	[40]
Lentil flour	0.5	2.1	49	0.86	23.1	—	[14]
	1	6.3	42	4.8	23	—	[14]

Table 5.
Mechanical properties, thermal degradation, and biodegradability of TPS and natural fiber-reinforced TPS.

method known as thermogravimetric analysis (TGA), where one of the test methods evaluates the temperature it takes to cause a 10% weight loss and records the differences among the different composites, with higher values indicating improved stability. Compared to TPS, the temperature at which 10% weight loss occurs increased from 192 to 229°C and 251°C corresponding to a 50 and 80% increase in the flax fiber fraction, respectively.

3.2.3 Biodegradation

Biodegradability is an integral process of biocomposites, which occurs as a result of microbial bacteria or fungi naturally assimilating the material structure and causing its degradation [16]. The main scheme followed to test the biodegradability rate is by measuring the percentage weight loss during a period of time. Generally, the higher the percentage of fibers, the lower the degradability rate due to the lower degradation rate of the fibers than starch-based polymers, as shown in **Table 5**. During the 1-week test period dictated to measure the rate of biodegradability, only 5% weight loss has occurred in the 80% flax fiber-reinforced TPS compared to the loss of 16% in the 40% flax fiber-reinforced TPS and 30% loss in pure TPS.

3.3 Design for a life approach

The design for a life approach assumes that the lifetime of a product can be estimated based on the rate of biodegradation, which depends on the material composition and service conditions. Data on biodegradation of various materials, such as that in **Table 5**, can be used to select the appropriate material for a given service environment and expected useful life of the product. The different mechanisms associated with the life cycle assessment of polymers is provided in [9], which provides a comprehensive overview of the needed knowledge for the determination of the useful life of the polymer. Moreover, Elsayed et al. presented biodegradation data related to flax fiber-reinforced starch composites, where the weight loss test was applied for long periods of time. The time needed for a 100% loss in weight was determined to be 6 weeks for TPS coinciding with a less than 40% reduction in the composite [28].

Successful application of the DFL approach entails a proper material selection process to be performed along with sufficient knowledge of the physical and chemical reactions associated with the proposed composites. This will provide the framework and foundation needed for choosing the applicable bio-based alternative and eventually help in controlling the service lifetime of the polymer by either accelerating the degradation process or stabilizing it depending on the application. The case study presented in Section 4.1 elaborates further on the use of material selection and substitution processes to aid in making the proper choice for applications in the automotive industry. The case study is adopted from material substitution cases reported by Farag and published in [16, 41].

4. Sustainable product design

4.1 Case study: the use of biodegradable composites in the automotive industry

Material selection processes are considered one of the most vital steps in the engineering and sustainable product design. This has become a necessary activity performed by automotive manufacturers and designers. Driven by the need to improve fuel efficiency, weight reduction has become a prime requirement. Accordingly, the fraction of lighter materials such as aluminum and plastic composites has progressively increased and substituted heavier steel alloys traditionally used. Other factors driving the search for alternatives are price, end-of- life vehicle legislation, and sustainability [16, 36]. Al-Oqla and Sapuan have emphasized on the importance of selecting the proper alternative biocomposite that meets all the requirements needed for environmental sustainability as well as compatibility to performance prerequisites [8]. Also, the authors added that considering the tremendous need and awareness of environmental issues, natural fiber-reinforced composites have become of major interest by researchers. Given their low density, good mechanical properties, renewability, and biodegradability, automotive interiors could be designed with high specific strength and stiffness properties, meeting design requirements while still meeting environmental criteria [37, 42].

4.1.1 Materials and composites for interior panels

Conventionally, polymers such as polyvinyl chloride (PVC) have been used for the interior panel structures [41]. Advantages such as easy processing and low cost have led to its extensive use in a wide variety of applications. However, it is a synthetic polymer with recycling and degradability issues making it an unfavorable choice. Subsequently, alternatives have been proposed in the literature with a recent review summarizing the selection criteria for biocomposites to be used in automotive structures [42]. **Figure 1** illustrates the use of hemp fibers in reinforcing polypropylene composites as a substitute in car doors.

Holbery and Houston have indicated that the use of bast fibers and specifically flax fibers presents a strong competition against E-glass fibers commonly used in composites implemented in automotive applications [43], where the specific strength for flax fiber is 1200 compared to 1275 for E-glass fibers [43]. Other natural fibers suggested for reinforcing plastics are hemp and jute. Their use has been reported in reinforcing PP replacing fiber glass-reinforced plastics in commercial Mercedes-Benz and Ford cars [44]. **Figure 2** demonstrates the use of flax fiber composites in different components of the Mercedes-Benz A-Class vehicle.

However, these composites are argued to be only partially biodegradable due to the synthetic matrices and hence are not an environmentally friendly option. A rather more sustainable option is the use of natural polymers such as starch reinforced by natural fibers. Nevertheless, this option has not been investigated in the literature in applications related to the automotive industry in spite of the benefits these composites offer, which range from the low energy needed for production to being renewable and biodegradable. The case study presented in Section 4.1.2 evaluates the use of natural fiber-reinforced starch as a potential candidate for substitution.

4.1.2 Performance indices and material requirements

For interior panels, the material requirements needed are lightweight and high stiffness. Cost and environmental considerations are other factors considered for the decision-making process. Thus, the material performance index (m) for a stiff light structural member is calculated based on the consideration that a panel is

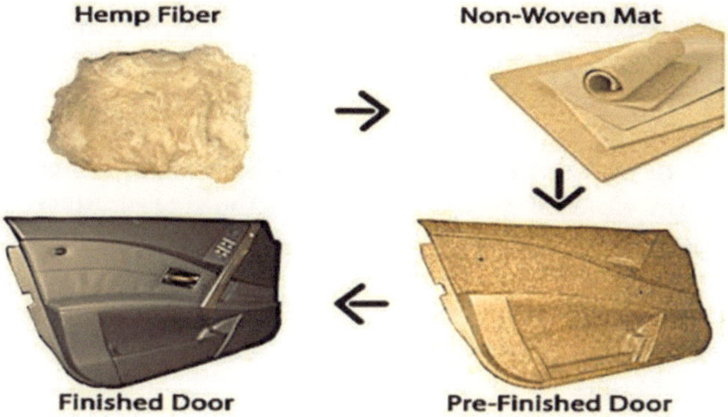

Figure 1.
Hemp fibers in vehicle doors [45].

Figure 2.
Mercedes-Benz A-Class vehicle components made of flax fiber composites [46].

rectangular of 100 cm in length (l), 50 cm width (b), and thickness (t) 3.7 mm for the PVC conventional material:

$$m = E^{1/3}/\rho \quad (1)$$

The mass (M) of the panel is

$$M = \rho tbl \quad (2)$$

The thickness is given by

$$t_n = t_o \left(E_o/E_n\right)^{1/3} \quad (3)$$

where t_n is thickness of alternative material, t_o is thickness of the PVC conventional material, E_o is elastic modulus of alternative material, and E_n is elastic modulus of the PVC conventional material.

Calculations for each performance index are presented in **Table 6** for the different candidate materials.

4.2 Cost of the panel

The total cost (C_t) of a panel is the summation of the cost of material, cost of manufacturing and finishing, cost over the entire life of the component (running cost), and cost of disposal and recycling.

4.2.1 Cost assumptions

The cost of the material in the panel is based on its weight and the price of material per unit weight. The manufacturing cost is roughly estimated based on the assumption that compression molding is used. According to Farag, the life of a car could be estimated to be 5 years, a total of 200,000 km traveled, $3/gal of fuel, and 8.62 km/L for a 1782 kg vehicle; the total fuel cost savings of the vehicle is approximately $6.6/kg. This amount can also be taken as the share in the running cost of a component weighing 1 kg over the entire life of the vehicle. Finally, the cost of disposal and recycling is estimated as being proportional to the weight of the panel and its material. Synthetic composites and matrices are difficult to dispose; thus the cost is assumed as $0.7/kg, while natural composites are relatively easier, and the cost of disposal is assumed to be $0.5/kg.

Finally, the cost of pure synthetic polymers is considered to be easy and estimated to be $0.3/kg., while biodegradable natural fibers are easiest to dispose of with a cost of 0.15/kg.

4.3 Environmental considerations

Motorcar weight reduction is considered as the most important factor in reducing the negative impact on the environment. This is related to the reduction in fuel consumption and the reduction in carbon dioxide emissions [47]. Therefore, this study assumes that the environmental impact of the panel is directly proportional to its weight.

4.4 Comparison of candidate materials using the compound objective function method

Table 7 gives the normalized values for each of the computed cost and weight of the panel. The performance index of a panel made of a given material is taken as

Material	E (GPa)	ρ (g/cc)	t (mm)	Cost of material /kg ($)	Cost elements ($)				Total cost of panel ($)	Weight of panel (Kg)
					Materials	Manufacturing	Running	Disposal		
PVC	2	1.3	3.7	1.375	3.3	2	15.84	0.72	21.86	2.4
PP and 40% GF	7.75	1.78	2.36	3.46	7.3	2	13.86	1.47	24.60	2.1
PP and 40% flax	4.65	1.34	2.8	1.04	2	2	12.4	0.94	17.34	1.88
Starch and 50% flax	4.5	1.4	2.82	0.34	0.7	2	13	0.3	16	1.97

Table 6.
Material properties and cost elements of candidate materials.

Material	Total cost of panel ($)	Normalized cost	Weight of panel as a measure of environmental impact	Normalized environmental impact	Performance index			
					Scenario 1	Rank	Scenario 2	Rank
PVC	21.86	73.2	2.4	78.33	75.8	3	90.9	3
PP and 40% GF	24.60	65	2.1	89.5	77.2	4	74.8	4
PP and 40% flax	17.34	92	1.88	100	96	2	95.2	2
Starch and 50% flax	16	100	1.97	95.4	97.7	1	98.16	1

Table 7.
Ranking of candidate materials.

the weighted sum product of the normalized values of its cost and environmental impact. Using the objective function method, two scenarios are considered: in the first scenario, the cost of the panel and its environmental impact are given equal weight, and in the second scenario, the cost of the panel is considered less important and is given a weight of 40%, and the environmental impact is considered more important and is given a weight of 60%. Both scenarios give the starch-flax composite the first rank. Its low cost more than compensates for its moderate weight.

5. Conclusion

Conventional plastics are designed with little consideration for their ultimate disposability or recyclability. Accordingly, this has led to the growing environmental awareness and notion toward the use of alternatives to petrochemical-based polymers. Given the ubiquity of plastic use in everyday life, substantial progress was made in the development of a reliable substitute, and in recent years, significant advancement was achieved in the production of alternative biodegradable materials based on renewable resources, which can offer equivalent functionality and physical properties similar to their petrochemical-based counterparts. Products that are based on such materials can be designed to biodegrade at the end of their useful life using the design for a life approach which entails that the material used in making a given product will not last long after the end of its useful life. However, the challenge is to design polymers to provide the required functionality during use and naturally degrade after. Consequently, this chapter has elucidated the advancement achieved by researchers in fabricating biodegradable alternatives from starch-based composites.

Abbreviations

T_g	glass transition temperature
T_m	melting temperature
UTS	ultimate tensile strength
E	Young's modulus
ε	elongation
ρ	density

Author details

Dina Fouad* and Mahmoud Farag
The American University in Cairo, Cairo, Egypt

*Address all correspondence to: dinafouad@aucegypt.edu

References

[1] Wróblewska-Krepsztul J, Rydzkowski T, Borowski G, Szczypiński M, Klepka T, Thakur VK. Recent progress in biodegradable polymers and nanocomposite-based packaging materials for sustainable environment. International Journal of Polymer Analysis and Characterization. 2018;**23**: 383-395. DOI: 10.1080/1023666x.2018. 1455382

[2] Ibrahim H, Mehanny S, Darwish L, Farag M. A comparative study on the mechanical and biodegradation characteristics of starch-based composites reinforced with different lignocellulosic fibers. Journal of Polymers and the Environment. 2018; **26**:2434-2447. DOI: 10.1007/ s10924-017-1143-x

[3] Spierling S, Knüpffer E, Behnsen H, Mudersbach M, Krieg H, Springer S, et al. Bio-based plastics—A review of environmental, social and economic impact assessments. Journal of Cleaner Production. 2018;**185**: 476-491. DOI: 10.1016/j.jclepro. 2018.03.014

[4] Arun Kumar K, Madhu Sudhanan S, Mahesh Kumar K, Ranjith Kumar G. A study on properties of natural fibers—A review. Internal Research Journal of Engineering and Technology. 2017;**4**: 1343-1350

[5] Sudesh K, Iwata T. Sustainability of biobased and biodegradable plastics. Clean. 2008;**36**:433-442. DOI: 10.1002/ clen.200700183

[6] Lu DR, Xiao CM, Xu SJ. Starch-based completely biodegradable polymer materials. ExpressPolymerLetters. 2009; **3**:366-375. DOI: 10.3144/ expresspolymlett.2009.46

[7] Vroman I, Tighzert L. Biodegradable polymers. Materials. 2009;**2**:307-344. DOI: 10.3390/ma2020307

[8] Al-Oqla FM, Sapuan S. Natural fiber reinforced polymer composites in industrial applications: Feasibility of date palm fibers for sustainable automotive industry. Journal of Cleaner Production. 2014;**66**:347-354. DOI: 10.1016/j.jclepro.2013.10.050

[9] Laycock B, Nikolić M, Colwell J, Gauthier E, Halley P, Bottle S, et al. Lifetime prediction of biodegradable polymers. Progress in Polymer Science. 2017;**71**:144-189. DOI: 10.1016/j.prog polymsci.2017.02.004

[10] Jia W, Gong RH, Hogg PJ. Poly (lactic acid) fibre reinforced biodegradable composites. Composites Part B: Engineering. 2014;**62**:104-112. DOI: 10.1016/j.compositesb.2014. 02.024

[11] Sharma VK, Sharma M. Plastics: Issues challenges and remediation. International Journal of Waste Resources. 2014;**04**;1-6. DOI: 10.4172/ 2252-5211.1000134

[12] Rochman C, Browne M, Halpern B, Hentschel B, Hoh E, Karapanagioti H, et al. Policy: Classify plastic waste as hazardous. Nature. 2013:169-171. Available from: https://www.nature. com/articles/494169a [Accessed: 17 June 2019]

[13] European Bioplastics. Bioplastics market data 2018 Report. 2018

[14] Ochoa-Yepes O, Medina-Jaramillo C, Guz L, Famá L. Biodegradable and edible starch composites with fiber-rich lentil flour to use as food packaging. Starch-Carbohydrate Polymer Composites. 2018;**70**(7–8):1700222. DOI: 10.1002/star.201700222

[15] Gironi F, Piemonte V. Bioplastics and petroleum-based plastics: Strengths and weaknesses. Energy Sources, Part A: Recovery, Utilization, and

Environmental Effect. 2011;**33**: 1949-1959. DOI: 10.1080/15567030903436830

[16] Farag MM. Design and manufacture of biodegradable products from renewable resources. In: Handbook of Composites from Renewable Materials; Scrivener Publishing-Wiley; 2017. pp. 111-131. DOI: 10.1002/9781119441632.ch23

[17] Gonzalez-Garcia S, Sanye-Mengual E, Llorach-Masana P, Feijoo G, Gabarrell X, Rieradevall J, et al. Sustainable design of packaging materials. In: Muthu SS, editors. Environmental Footprints of Packaging. Singapore: Springer; 2016. P. 23–46.

[18] Vieira AC, Guedes RM, Tita V. Considerations for the design of polymeric biodegradable products. Journal of Polymer Engineering. 2013;**33**:293-302. DOI: 10.1515/polyeng-2012-0150

[19] Anne B. Environmental-friendly biodegradable polymers and composites. Integrated Waste Management. 2011;**1**:343-364. DOI: 10.5772/16541

[20] Cruz-Romero M. Crop-based biodegradable packaging and its environmental implications. CAB Reviews: Perspectives in Agriculture, Veterinary Science, Nutrition and Natural Resources. 2008;**3**(074):1-22. DOI: 10.1079/pavsnnr20083074

[21] Doppalapudi S, Jain A, Khan W, Domb A. Biodegradable polymers—An overview. Polymers for Advanced Technologies. 2014:427-435. DOI: 10.1002/pat.3305

[22] Muthuraj R, Misra M, & Mohanty A. Studies on mechanical, thermal, and morphological characteristics of biocomposites from biodegradable polymer blends and natural fibers. In: Biocomposites. Design and Mechanical Performance. Woodhead Publishing; 2015:93-140. DOI:10.1016/b978-1-78242-373-7.00014-7

[23] Okada M. Chemical syntheses of biodegradable polymers. Progress in Polymer Science. 2002;**27**: 87-133. DOI: 10.1016/s0079-6700(01)00039-9

[24] Soni S, Gupta H, Kumar N, Nishad D, Mittal G, Bhatnagar A. Biodegradable biomaterials. Recent Patents in Biomedical Engineering. 2010;**3**:30-40

[25] Chen B, Evans J, Greenwell H, Boulet P, Coveney P, Bowden A, et al. A critical appraisal of polymer–clay nanocomposites. Chemical Society Reviews. 2007:568-594. DOI: 10.1039/b702653f

[26] Rezwan K, Chen Q, Blaker J, Boccaccini AR. Biodegradable and bioactive porous polymer/inorganic composite scaffolds for bone tissue engineering. Biomaterials. 2006;**27**: 3413-3431. DOI: 10.1016/j.biomaterials.2006.01.039

[27] Gross R, Kaira B. Biodegradable polymers for the environment. Science. 2002;**297**:803-806

[28] Elsayed H, Farag M, Megahed H, Mehanny S. Influence of flax fibers on properties of starch-based composites. Design, Materials and Manufacturing, Parts A, B, and C. 2013;**3**:1397-1408. DOI: 10.1115/imece2012-89628

[29] Ibrahim H, Farag M, Megahed H, Mehanny S. Characteristics of starch-based biodegradable composites reinforced with date palm and flax fibers. Carbohydrate Polymers. 2014; **101**:11-19. DOI: 10.1016/j.carbpol.2013.08.051

[30] Mehanny S, Farag M, Rashad RM, Elsayed H. Fabrication and characterization of starch based bagasse

fiber composite. Design, Materials and Manufacturing, Parts A, B, and C. 2012; **3**:1345-1353. DOI: 10.1115/imece2012-86265

[31] Gilillan W. Developing Starch-based polymer composites [thesis]. Queensland university of Technology; 2015

[32] Bikiaris D, Prinos J, Koutsopoulos K, Vouroutzis N, Pavlidou E, Frangis N, et al. LDPE/plasticized starch blends containing PE-g-MA copolymer as compatibilizer. Polymer Degradation and Stability. 1998;**59**:287-291. DOI: 10.1016/s0141-3910(97)00126-2

[33] Reis MO, Olivato JB, Bilck AP, Zanela J, Grossmann MV, Yamashita F. Biodegradable trays of thermoplastic starch/poly (lactic acid) coated with beeswax. Industrial Crops and Products. 2018;**112**:481-487. DOI: 10.1016/j.indcrop.2017.12.045

[34] Lalit R, Mayank P, Ankur K. Natural fibers and biopolymers characterization: A future potential composite material. Strojnícky Casopis—Journal of Mechanical Engineering. 2018;**68**:33-50. DOI: 10.2478/scjme-2018-0004

[35] Reddy MM, Vivekanandhan S, Misra M, Bhatia SK, Mohanty AK. Biobased plastics and bionanocomposites: Current status and future opportunities. Progress in PolymerScience. 2013;**38**(10–11): 1653-1689. DOI: 10.1016/j.progpolymsci.2013.05.006

[36] Chellaperumal D, Karthikeyan N, Nandhakumar P, Ekanthamoorthy J. Fabrication and mechanical testing on natural fiber composite. International Research Journal of Engineering and Technology. 2019;**6**:1095-1098

[37] Ku H, Wang H, Pattarachaiyakoop N, Trada M. A review on the tensile properties of natural fiber reinforced polymer composites. Composites Part B: Engineering. 2011;**42**:856-873. DOI: 10.1016/j.compositesb.2011.01.010

[38] Gironès J, López J, Mutjé P, Carvalho A, Curvelo A, Vilaseca F. Natural fiber-reinforced thermoplastic starch composites obtained by melt processing. Composites Science and Technology. 2012;**72**:858-863. DOI: 10.1016/j.compscitech.2012.02.019

[39] Boudjema H, Bendaikha H. Studies on the properties of cellulose fibers reinforced thermoplastic starch composites. Journal of Chemistry and materials research. 2015;**3**:21-25

[40] Torres FG, Arroyo OH, Gomez C. Processing and mechanical properties of natural fiber reinforced thermoplastic starch biocomposites. Journal of Thermoplastic Composite Materials. 2007;**20**(2):207-223. DOI: 10.1177/0892705707073945

[41] Farag MM. Materials and Process Selection for Engineering Design. Boca Raton, FL: CRC Press; 2014

[42] Noryani M, Sapuan SM, Mastura MT, Zuhri MY, Zainudin ES. Material selection criteria for natural fibre composite in automotive component: A review. IOP Conference Series: Materials Science and Engineering. 2018;**368**:012002. DOI: 10.1088/1757-899x/368/1/012002

[43] Holbery J, Houston D. Natural-fiber-reinforced polymer composites in automotive applications. Journal of Metals. 2006;**58**(11):80-86

[44] Sapuan S, Kho J, Zainudin E, Leman Z, Ahmed Ali B, Hambali A. Materials selection for natural fiber reinforced polymer composites using analytical hierarchy process. Indian Journal of Engineering and Materials Sciences. 2011;**18**:255-267

[45] Pecas P, Carvalho H, Leite M. Natural fibre composites and their applications: A review. Journal of

Composites Science. 2018;**2**:1-20. DOI: 10.3390/jcs2040066

[46] March G. Next step for automotive materials. Materials Today. 2003;**6**: 36-43. DOI: 10.1016/S1369-7021(03) 00429-2

[47] Fogorasi M, Barbu I. The potential of natural fibres for automotive sector—Review. IOP Conference Series: Materials Science and Engineering. 2017;**252**:012044. DOI: 10.1088/1757-899x/252/1/012044

Chapter 4

Integrating Sustainability in the Strategic Stage of an Innovation Process: A Design Brief Perspective

Kristel Dewulf

Abstract

This book chapter explores the uptake of environmental sustainability in the front end of an innovation process, and the outcome of this stage: the design brief. The study is based on a content analysis of 80 design briefs from Belgian enterprises, a focus group with representatives from 14 Belgian companies and an in-depth interview with two Belgian chief executives. The results show an overview of the most and least used environmental sustainability strategies in the design brief and demonstrate a remarkable difference in uptake between large enterprises and small and medium-sized enterprises. Findings show that companies often deal with sustainability on a hidden and decomposed level. Furthermore, a generic model for the design brief process is presented with the different entry points for sustainability. Crucial factors for integrating environmental sustainability in the design brief are discussed in the last section. The paper concludes with a recommendation to integrate ecodesign targets in the design brief and discussing them with the decision makers in all the stages of the design briefing process. With no environmental commitment in the design brief towards the final product, no time, budgets, and staff will be allocated on this subject during the operational stage.

Keywords: design brief, sustainable product innovation, ecodesign, front end innovation, strategic innovation management, new product development

1. Introduction

The very early phase in the innovation process, the so-called front end of innovation (FEI), is often described as being the root of success for any company hoping to compete on the basis of innovation [1]. It is the phase with the largest impact on the end result of the project [1, 2] and the highest payback to one's investments [2]. It is in this phase that companies set their targets and determine which products will pass to further development. The outcome of that process is usually reflected in the design brief [3]. The decisions made in the front end (FE) and the design brief influence all the later phases of the innovation process.

This chapter argues that the uptake of sustainability in design projects would be far more effective if sustainability aspects are written down in the design brief, providing guidance to the design, engineering, marketing, and management team.

Till now, there is limited understanding of how to best bring environmental considerations into the design brief. FEI is a hot research topic, but its relationship with the design briefing process in the FE and design for sustainability has received much less attention in literature and practice. There are a number of tools available to guide designers, engineers, and managers in the design process after the specifications of the product or service are already set, but methods supporting goal finding for sustainable innovations are rare [4].

This book chapter describes a first exploratory study to fill this gap. The research aims at gaining understanding on how environmental sustainability is integrated in the strategic stage of an innovation process, the so-called FE, and the design briefing process. The first part is based on a literature review and elaborates the FE and what is meant by a design brief. The second part looks to the current practice of integrating environmental considerations in a design brief and the FE. Different research results are described and discussed in the penultimate section. The chapter concludes with recommendations for future research.

2. Literature review

2.1 The strategic stage in an innovation process

Innovation projects in industry generally move along three major activity domains: The pre-development activities where future products are defined and decided on (the FE), the development activities where these products are actually developed (the New Product Development phase (NPD)), and the launching or commercialization activities where these newly developed products are brought to the market [2].

The strategic stage in the product innovation process is the stage where product strategy formulation, opportunity identification, idea generation, idea selection, and concept development take place and decisions about new product development are taken [2]. According to Kim and Wilemon [5], the FE starts when an opportunity is considered valuable of deeper exploration and ends when a company concludes to invest significant resources to the development of the idea and launch the project. The FE includes product strategy formulation and communication, opportunity identification and assessment, idea generation, product definition, project planning, and early executive reviews, which typically precede detailed design and development of a new product [5].

According to Crawford and Di Benedetto [6], the FE process answers these five essential questions: what, why, who, when, and how.

- What: project description.
- Why: the strategy behind the project.
- Who: the necessary human resources.
- When: project timing.
- How: describes all the product requirements regarding the new development.

Similar to Jacoby [7], the FE phase in this chapter is defined as "all initial innovation activities, prior to the phase where real new product development starts".

The FE has typical characteristics; less information is known, there is a high degree of freedom, and the cost for changes is low. Later in the process more information is disposable, but changes will have a higher price [8, 9]. It is under these uncertain circumstances that decisions are made and the design brief is formulated in the FE.

2.2 Importance of the strategic stage

The root of success of innovations can be found in the activities done in the FE [1]. The first phases in the innovation process have the largest impact on the end result of the project and the highest payback to one's investments [2]. Quality, costs, and timings are mostly set during the FE. Important decisions are made here, and will follow the product through the project [8]. Choices made in the FE influence the success of a product on a great extent [7].

The impact decisions can have on the end product diminishes during the project. FE decisions can have an impact on the whole product, whereas NPD decisions only have an impact on partial aspects as the team has to take into account earlier choices [7]. Koen and Bertels [10] also highlight this path-dependency in an innovation process. The most significant benefits can be achieved through improvements of the FE activities [11]. Also a study by Koen et al. [2] states that the FE presents one of the greatest opportunities for improving the overall innovation process. Verworn [12] concludes that a better understanding of the FE increases the success rate of the NPD process.

2.3 The design brief

The outcome of the FE is usually reflected in the design brief [3] and can be seen as the point of transfer between the FE and the NPD [13, 14]. It is a written description, an agreement or contract of a project that requires some form of design between the parties involved in the project [3]. Often, the design brief serves as a point of transfer between different professionals, where the project is handed over from marketing to design, or from a product manager to an external design agency. The role of a design brief is to foresee the base of the design process in the form of a practical paper that reflects the final product's attributes [15].

A good design brief tries to obtain the knowledge of the design and management team, the expert, and the user and the buyer. Formulating a design brief is a creative, iterative, and interactive process and is best developed in partnership [3]. Successful briefing is about clear and comprehensive communication and how information is structured [16]. It is important that the brief contains all the information and data necessary for every stakeholder [3].

A variety of terminologies are used for design assignments in literature and practice. People may refer to the design brief as new product development brief, creative brief, project brief, project sheet, innovation brief, or marketing brief [3, 16, 17]. Design is a broad term, with a variety of design disciplines (e.g. industrial design, package design, and communication design). Each discipline requires different information in a truly useful design brief [3]. The focus of this paper is "Industrial Design Briefs", briefs for the design of a new consumer product or a product-service system.

2.4 Elements of the design brief

While in practice the structure of design briefs may vary, Phillips [3] describes eight elements that he seems essential for a good design brief, as explained in **Table 1**. The table also represents a schematic overview of a good design brief.

Design brief element	Description
(1) Project overview and background	The executive summary of the project. It clearly articulates the scope, the business needs, objectives, desired outcomes and ownership of the project in terms non-design business partners can understand. It contains all the essential information: what, why and who of the project
(2) Category review	The industries in which the product or service is involved, including market positioning, brand and business strategy and significant trends
(3) Target audience review	The target audience(s) description, as completely as possible
(4) Company portfolio	Decription of the company and its activities
(5) Business objectives and design strategy	This section provides a clear understanding of how design strategy is matched to business objectives
(6) Project scope, timeline and budget	Project scope and time-budget requirements in every phase of the project
(7) Research questions	Unanswered questions that are critical to the success of the design project
(8) Appendix	Optional section that might contain documents that summarize research data, competitive analyses and visual material

Table 1.
Essential elements of a good design brief, according to Phillips (2004).

3. Research approach

A variety of authors have recommended to focus on the FEI in the eco-innovation literature in the last decade [17–27] However, with few exceptions [17, 28], little attention is given in the literature regarding the design brief process in the FE in relation to sustainable product innovation. There is still insufficient comprehension of this important matter.

Previous research by Boks [29] has identified the use of environmental checkpoints, reviews, milestones, and roadmaps as an important success factor regarding the integration of sustainability considerations in the early stages of the product development process. As the design brief can be seen as a roadmap and project-tracking tool that defines the various steps that will be followed [3], the assumption was made that the design brief can play an important role in achieving this.

The research in this chapter describes a first exploratory study to fill this gap. It aims at gaining understanding on how environmental sustainability is integrated in the design brief in the FEI. Based on the insights from the literature and previous explorative studies [19, 27–29], the research questions can be formulated in order to address the research objective as follows:

How frequently do companies add environmental sustainability in their design briefing? (1a).

If they do so, which ecodesign principles are requested in the design brief? (1b).

The answer to question "1a" will teach us how frequently environmental sustainability elements in the design brief are mentioned in the data sample. Question "1b" on the other hand will help us to understand in more detail what ecodesign principles are used. Seeking a plausible explanation for the reason as to why certain ecodesign principles are more frequently mentioned in the design briefs than others can only be answered by firstly regarding the design briefing process in the FE, as there is little research to date on this topic. By having insight into this design briefing process, it would be possible to define the entry points for sustainability and key factors for integrating sustainability in design brief. As such, the next research questions were formulated:

How is the design briefing process established in the FE of an innovation process? (2a).

Where are the entry points for sustainability? (2b).

And what are success factors for integrating environmental sustainability in the design brief? (2c).

Three exploratory studies have been conducted with a focus on Belgium SMEs and large enterprises within various industries to answer the questions above. In order to enhance the credibility in the findings, a between-method triangulation was chosen by involving three research methods.

Because relatively little research has been conducted regarding environmental parameters in a design brief and the design brief process, preference went to a qualitative approach, since the focus lies on a small sample to try to look at a range of interconnected processes and causes.

In the first study, a content analysis of 80 industrial design briefs from 62 Belgium-based companies was used as research methodology. The found ecodesign elements in the design briefs were typified and assigned according to the Ecodesign Strategy Wheel [30].

In study 2, a focus group was organized with representatives from 14 Belgium SMEs and large enterprises, in combination with a double in-depth interview with two Belgian chief executive officers (CEOs). These participatory methods were chosen for some particularly reasons; it allows participants to question each other and to elaborate upon their answers. The participants can develop and influence each other's ideas and opinions in the course of discussion. It is also useful when there is a desire to learn more about consensus on a topic and when one is interested in complex motivations [1]. The research procedure is explained in detail in the next sections.

Sleeswijk Visser et al. [31] have shown the relationships between the various forms of data gathering and their ability to access different types of knowledge. Corresponding with these insights, a mix of different research techniques are used in this study in order to get access to the following levels of knowledge; explicit (interview), observable (content analysis of the 80 design briefs), and tacit/latent (focus group).

4. Study 1: The analysis of 80 design briefs

4.1 Research approach for the analysis of 80 design briefs

A design brief is an essential communication paper between the company and the design bureau. It generally gives an accurate insight in the sustainability ambition a company has for their future product or service. But how often do corporations add ecodesign topics in their assignment? If they do so, which ecodesign principles are requested in the design brief?

To answer these questions, a content analysis of 80 industrial design briefs from 62 Belgium-based companies was used as research methodology. Among those companies, 50% was categorized as small and medium-sized enterprises (SMEs), 41% as large enterprises, while the other 9% were classified as "other" (knowledge institutions, government, universities, or associations of industry-specific institutions). The European definition (EU, 2003) was used to categorize the SME businesses. According to this definition, the main factors determining whether a company is an SME are number of employees (<250) and either turnover (≤€ 50 m) or balance sheet total (≤€ 43 m).

In all the cases, the companies firstly contact the design team with a request for proposal. This document becomes a design brief after reviewing and discussing the problems and needs with both parties.

The level of innovation proposed in the design briefs in this study varies from average to high. The design briefs can be categorized as open briefs, where the outcome of the project is not yet clearly defined and the product parameters are still flexible. All assignments covered physical, tangible products. More than 95% represents "end products", while the remaining 5% exists out of semi-finished product.

To get a good view on the daily practice and to close out socially desirable behavior, no one was informed in advance about this study.

4.2 Content analysis of 80 design briefs

Different methods supporting the analysis of a product's impact on the environment can be found in literature and practice. A method was needed to analyze the design briefs that could be applied in the first stage of a product design process, with a general product in mind. Such a method was found in the Ecodesign Strategy Wheel, also called Lifecycle Design Strategies (LiDs) [30]. The method helps to select and communicate strategies to minimize the environmental impact of a design.

The Ecodesign Strategy Wheel provides eight EcoDesign strategies that can be considered systematically, as presented in the **Table 2**. Strategy 0 is either "strategic", working on the product concept level, while the other strategies represents the product life cycle and relates to the product component, structure, and product level. The strategies are divided into 32 sub-strategies and correspond to possible solutions to improve the environmental profile of a product.

The found ecodesign elements in the design briefs were typified and assigned according to the strategies, and indicated as being quantitatively or qualitatively. No distinction is made in the analysis between projects that are initiated with the intention of doing something sustainable vs. projects without a specific sustainability focus.

4.3 Limitations of the study

The sustainability of the final product cannot be deducted from the design brief, as earlier research showed [17]. The incorporation of sustainability in a design brief does not guarantee results. Several organizational issues could function either as success or failure factors for the entire process. The opposite is also possible; in the case that the design brief does not express any wish or desire for sustainability, it is still possible that the design team may bring sustainability later in the project [28] when new insights are obtained during the innovation process.

Secondly, project leaders and design team members can strongly influence the final outcome of the design brief. Ecodesign push and pull mechanisms can show up in discussions with the company and the design team and often influence the final content of a design brief.

At last, diverse sectors were covered in the design briefs, varying from the electric and electronic industry, lighting, furniture, medical equipment, building, engineering, technology, and polymer industry. This implies a large variety in end products and in terms of production techniques, materials, end-users, market volumes, product function, etc. Therefore, the conclusions to this study can only be indicative.

Lifecycle design strategies		
Product level	Strategy	Sub-strategy
Product concept level	New concept development	0.1 Dematerialization
		0.2 Shared use of the product
		0.3 Integrations of functions
		0.4 Functional optimization of product
Product component level	Selection of low impact materials	1.1 Cleaner materials
		1.2 Renewable materials
		1.3 Lower energy content materials
		1.4 Recycled materials
		1.5 Recyclable materials
	Reduction of material usage	2.1 Reduction of weight
		2.2 Reduction in (transport) volume
Product structure level	Optimization of production techniques	3.1 Alternative production techniques
		3.2 Fewer production steps
		3.3 Lower/cleaner energy consumption
		3.4 Less production waste
		3.5 Fewer/cleaner production consumables
	Optimization of distribution system	4.1 Less/cleaner/reusable packaging
		4.2 Energy-efficient transport mode
		4.3 Energy-efficient logistics
	Reduction of impact during use	5.1 Lower energy consumption
		5.2 Cleaner energy source
		5.3 Fewer consumables needed
		5.4 No waste of energy/consumables
Product system level	Optimization of initial lifetime	6.1 Reliability and durability
		6.2 Easier maintenance and repair
		6.3 Modular product structure
		6.4 Classic design
		6.5 Stronger product-user relation
	Optimization of end-of-life system	7.1 Reuse of product
		7.2 Remanufacturing/refurbishing
		7.3 Recycling of materials
		7.4 Saver incineration

Table 2.
Lifecycle design strategies and sub-strategies according to the product level [30].

4.4 Results of the design brief analysis

4.4.1 Overall uptake

The content analyses of the 80 design briefs have shown that the uptake of environmental considerations into design briefs is limited. There was a remarkable difference between large enterprises, SMEs and enterprises categorized as "other" (knowledge institutions, government, universities. or associations of

Type of company	Uptake (%)
SMEs	43
Large enterprises	64
Other	86

Table 3.
The uptake of ecodesign elements in the analyzed design briefs, split up by company type.

industry-specific institutions). SMEs scored poorly: in 43% of all the SME design briefs, there was an ecodesign component found. The results are presented in **Table 3**.

4.4.2 Uptake on product level

Figure 1 gives an insight into the environmental profile of the design briefs on the different product levels. A distinction is made between SMEs, large enterprises, and "other" companies, demonstrated in the three bars. Large enterprises exceed the SMEs in all the product levels.

4.4.3 Uptake on strategy level

The most frequently used strategies in the design briefs are presented in **Figure 2**. As one can see, most of the strategies fluctuated between 0 and 18%, with an exception for the strategy "optimization of initial lifetime". This strategy was most popular for all the company types.

4.4.4 Uptake on sub-strategy level

The results on sub-strategy level are presented in **Figures 3** and **4**. Overall, the most popular sub-strategy was reliability and durability (6.1) followed by modular

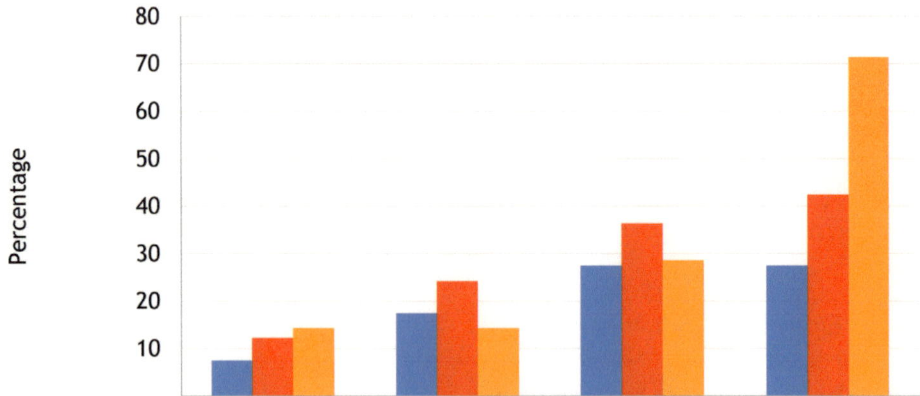

	Product Concept Level	Product Component Level	Product Structure Level	Product System Level
SMEs	8	18	28	28
Multinational	12	24	36	42
Other	14	14	29	71

Figure 1.
The integration of ecodesign in the analyzed design briefs on the different product levels (in percentage).

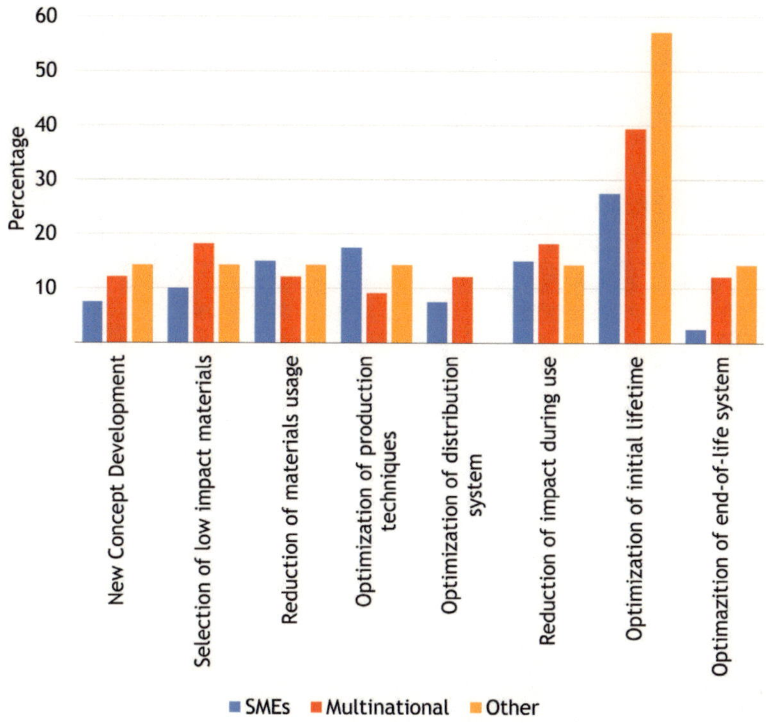

Figure 2.
Environmental profile of the design briefs on strategy level (according to [30]).

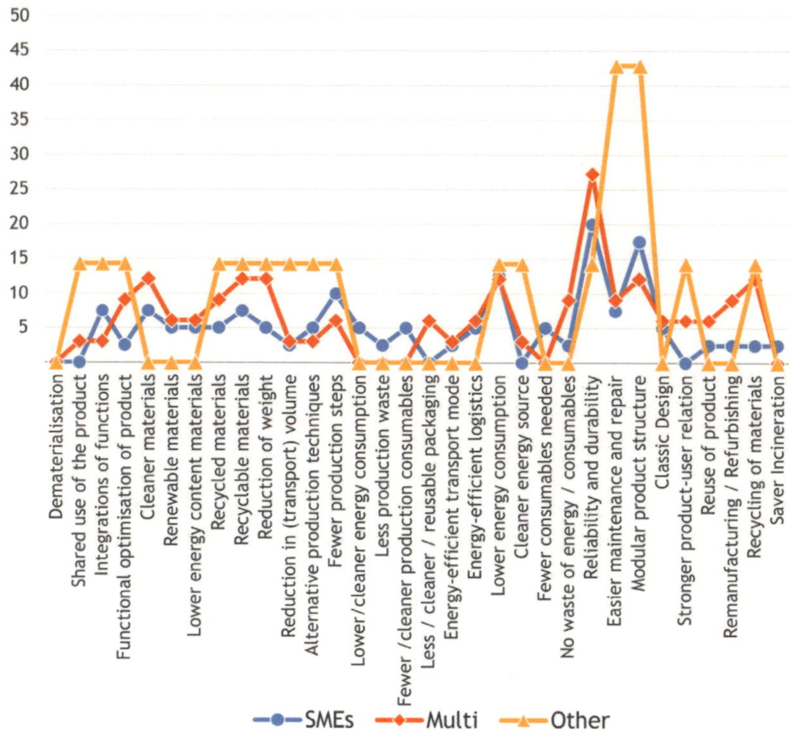

Figure 3.
Percentage of design briefs where a particular sub-strategy occurred, split up by company type.

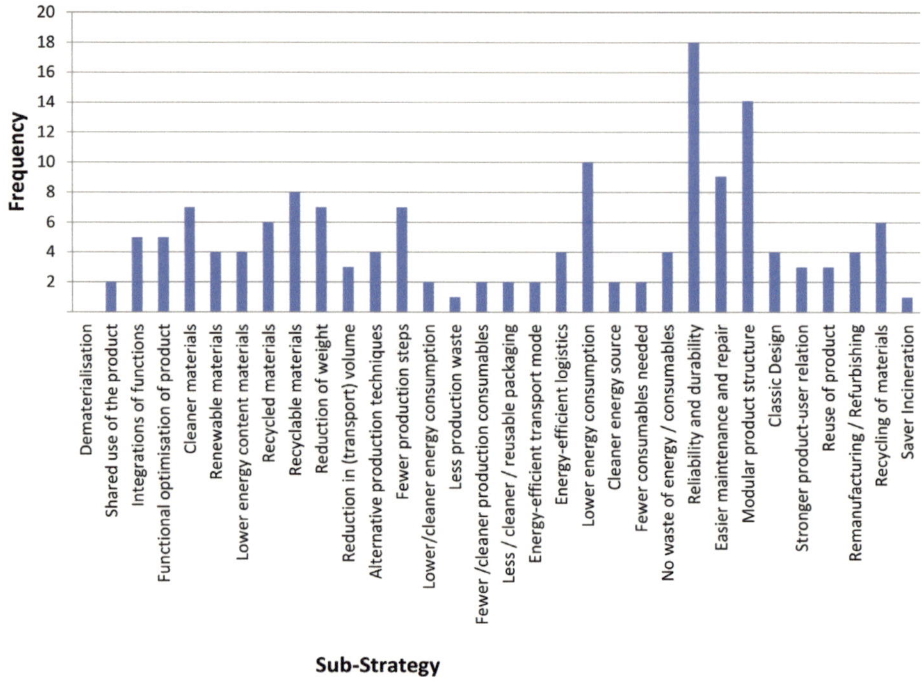

Figure 4.
The frequency of found sub-strategies in the 80 design briefs.

product structure (6.3) and lower energy consumption (5.1). Sub-strategy dematerialization (0.1) was not found once. The least found sub-strategies were less production waste (3.4), saver incineration (7.4), lower/cleaner energy consumption of production techniques (3.3), fewer/cleaner production consumables (3.5), less/cleaner reusable packing (4.1), and energy-efficient logistics (4.3). On average, in each design brief 1.6 sub-strategies were detected.

5. Study 2: focus group and interview

5.1 Research approach

Is there an explanation as to why certain design briefs more frequently mentioned environmental sustainability elements in the design briefs than others? And furthermore, is there a reason as to why certain sub-strategies are more frequently mentioned in the design briefs than others?

Before one can answer these questions, one must first have a clear view on the design brief process in the FEI. The assumption was that by having insight into this design briefing process, it would be possible to define possible entry points for sustainability. To give an answer to these questions, the following study was carried out.

First of all, a focus group was organized with representatives from 14 Belgium SMEs and large enterprises. There were two major criteria to participate in the focus group: being located in Belgium and having an active product development department, either by an in-house design team or in collaboration with an external design agency. All persons volunteered on the focus group after a call for participation. Background of the participants ranged between senior management, project management, product design, engineering, and research and development, as shown in appendix A. The focus group was organized as an interactive group setting where participants were divided in teams of four and five and lasted 1.5 h. Each team could make notes and schemes on big sheets of paper in order to visualize their thoughts to the other team members.

The following topics were discussed:

In the first part, the participants were asked to visualize and discuss their innovation process with the team members. Special attention was given to the FE of the process and the flow of the design brief. In the second part, the teams shared their experiences in integrating sustainability in the early stages of a design process and the design brief. Findings of the teams were summarized and presented plenary to all the participants of the focus group. In the end, the presented results were discussed with the whole group.

The focus group session and final presentation were recorded with notes and partly with audio. They were transcribed chronologically by use of sentences, key words, and statements. The sheets of paper were analyzed and summarized.

In addition to the focus group session, a double semi-structured in-depth interview with two CEOs from two Belgian SMEs was conducted to further clarify issues. The interview was carried out at the office of one of the CEOs, lasted 1 hour and a half, and was recorded both with audio and notes. The two CEOs were selected on their expertise. Both have a background in industrial product design; one has a specialization in Front End Innovation (>10 years of experience), the other in Environmental Sustainability and Sustainable Business Models (>15 years of experience). None of them has set any of the briefs that were analyzed in 4.1. They are both active in different sectors in product design and consultancy, and do not see each other as competitors. The idea of doing a double interview, instead of separately, came from them, to create a certain dynamic.

The two CEOs were interviewed about their FE innovation process, their experiences in design briefing, and the integration of sustainability in both of them. The interview questions can be found in appendix B of this chapter.

5.2 Findings

This section presents the main results based on the focus group and the interview.

5.2.1 Design briefing process and entry points for sustainability

As many people believe, the design brief is not a single document. Though processes differ from company to company, a multi-step design briefing process at the FE was found at all innovation processes of the participants. During this briefing process, different documents jump back and forward between different people and departments in the company.

With these insights, a new generic model of the design briefing process is presented in **Figure 5**. It shows the various stages of a design brief at the FEI. Although not all companies had such a formal organized process, there was a consensus in the group on the different stages and activities. The process is presented linear, but with different feedback loops. The model does not represent actual time frames. An earlier version of this model is described in another article of the authors of this chapter [20] as a result of a preliminary explorative study.

The various stages of a design brief document are explained in **Table 4**. There is no one-size-fits-all answer to the question "who is involved in the different stages of the design brief process". Different companies manage innovation in the FE in different ways. Also the decision maker(s) and the decision making process vary from company to company and are project-dependent. The people involved in the

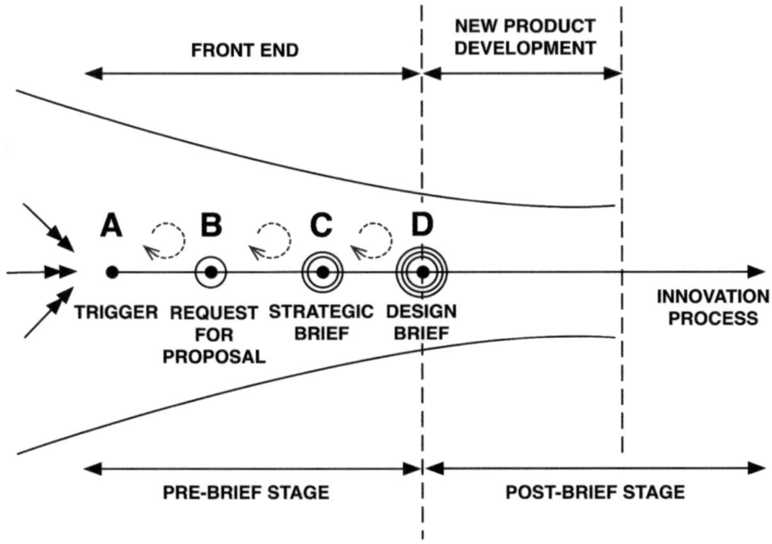

Figure 5.
The various stages of a design brief in an innovation process.

	Description
(A) Trigger	This is the start of the briefing process. Someone in the company defines a business objective or need. It can be a "gut" feeling based on previous experience, a recommendation from a stakeholder, or a result of research.
(B) Request for proposal	This document is sometimes also called "statement of need". It contains some basic information about the business objective or need and goes to the management level in the organization where the team can decide whether it is worth pursuing, mostly decided after conducting a feasibility study.
(C) Strategic brief	When the projects get a "go", the request for proposal will be upgraded into a "*strategic brief*" (C). This document is usually created for the in-house design team or for an external design agency.
(D) Design brief	Here the design brief is developed and written, usually in co-creation with the in-house design team or an external design agency after considerable thought and discussion about the project.

Table 4.
The various stages of the design brief.

briefing process can be a fixed team, an ad-hoc composed team, or it can even be one person where the entire FE is in his head.

After the interview it became clear that the converging funnel model, as presented in **Figure 5**, does not reflect the daily reality in many companies. A convergent and divergent stage, as well in the strategic stage as in the operational stage is a more common practice. The British Design Council has described such a model called the "Double Diamond" design process model [32]. Divided into four distinct phases, discover, define, develop, and deliver, it maps the divergent and convergent stages of the design process.

The design briefing process, shown in **Figure 5**, was revised and adapted based on this double diamond model. The result is presented in **Figure 6**.

The different stages in the design briefing process as presented in **Figures 5** and **6**, also mark the different entry points for sustainability. The earlier in the process, the more room there is for improvement.

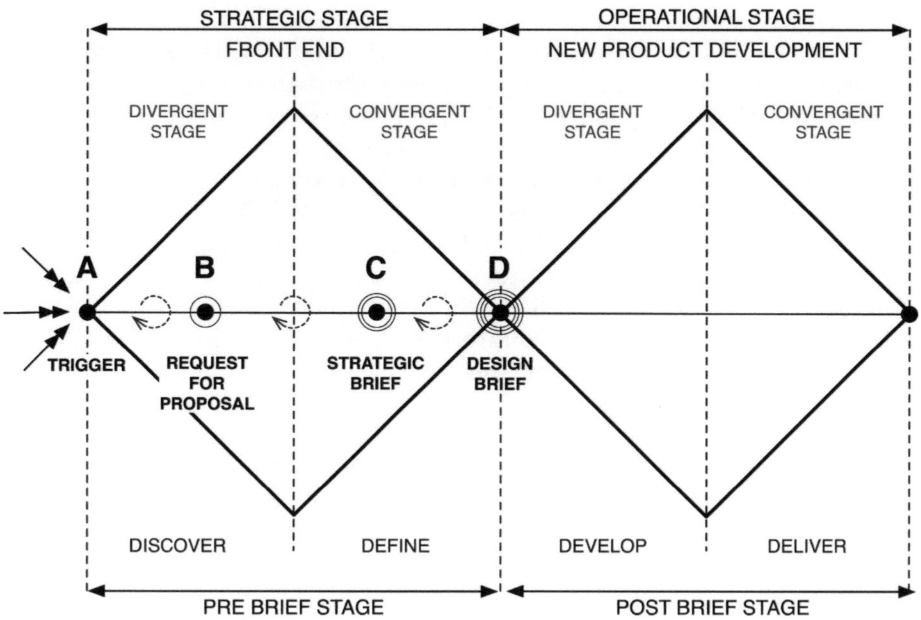

Figure 6.
The design briefing process integrated into the double diamond design process model [32].

5.2.2 Crucial factors for integrating sustainability in the design brief

All participants agreed on three crucial factors for integrating sustainability in the design brief. Firstly, the integration of sustainability in the FE and the design brief largely depends on who is involved in the design briefing process and who is making the decisions in the FE. People are dominant factors; they make decisions, not tools or methods. They determine what will be written down in the design brief and will set, whether or not, the environmental sustainability goals.

A second crucial factor is the commitment of the CEO and the management team. A sustainable product has to be embedded in a strategic sustainable framework, set up in the FE and translated in a design brief. These decisions cannot be taken down the chain; bottom-up does not work on innovation process level for sustainable product innovation without an engagement from the management team.

Thirdly, there are basically two reasons why the decision makers integrate environmental sustainability in the design brief; either they do it because they see business opportunities (market demand, cost reduction, product differentiation, marketing...) or they do it because it is required (legislation, retailer demands...). These drivers are also mentioned in the ecodesign literature [30]. The decision makers need to have a clear view on the business opportunities, needs, risks, and costs of sustainable product innovation in the FE. As long as these topics are not obvious, it will be difficult to convince the stakeholders, and to adapt it in the design brief. Cynically, due to the characteristics of the FE phase as explained in 2.1, it is a very tough exercise, as one has to make decisions in uncertain conditions.

6. Conclusions

This book chapter aims to contribute to the FE of eco-innovation literature. The main research question was whether companies mention certain ecodesign principles as distinguished in the Ecodesign Strategy Wheel [30] in the design brief, and whether the design brief process or the existence of specific elements in this process can explain this. The study is based on a content analysis of 80 design briefs from Belgian SMEs and large enterprises, a focus group with representatives from 14 Belgian companies and a double semi-structured in-depth interview with two Belgian CEOs.

This study has shown a remarkable difference between large enterprises, SMEs and companies categorized as "other" (knowledge institutions, government, universities, or associations of (industry-specific) institutions) in uptake of environmental considerations in Belgian design briefs. SMEs scored poorly in this study: in 57% of all the SME design briefs, there was no ecodesign component found.

Large enterprises exceeded the SMEs on all the product levels. A possible explanation for this can be found in the literature. Bocken et al. [18] pointed out that eco-innovation in the FE can be more easily mastered in big, resourceful companies. On the other hand, larger companies may have difficulties in allowing the eco-innovation process to be open, informal, and creative, aspects which contributed positively to the success of novel eco-innovations [18]. According to van Hemel et al. [30], larger companies are subject to more and stronger stimuli to take their responsibility towards green products than SMEs. They receive more media attention and are more vulnerable for criticism of external stakeholders.

Integration of ecodesign in the design brief was most frequently found on product system level. The most popular sub-strategy was "reliability and durability", followed by modular product structure and lower energy consumption. Sub-strategy dematerialization was not found once. The least found sub-strategies were less production waste, saver incineration, lower/cleaner energy consumption of production techniques, fewer/cleaner production consumables, and less/cleaner reusable packing and energy-efficient logistics. This confirms the findings of [28] as the result an explorative study on in-depth interviews with five major Dutch design agencies on how design agencies deal with sustainability issues in the FE; sustainability often appears to be dealt with on a decomposed level, with a focus on, that is, material reduction, or energy efficiency, and not on the holistic concept of sustainability.

Some Lifecycle Design Strategies can be categorized as "hidden sustainability", for instance reliability, durability, modular product structure, easier maintenance and repair... In some cases, where no explicit request for ecodesign was made in the design brief, still many Lifecycle Design Strategies were found. Often cost optimization, risk management, safety management, distribution planning, product warranty... were the driver, with a more sustainable project as a "side effect". This also proves the findings of Storacker [28]; agencies try to make sustainable "wise choices" in design, even if this is not something they necessarily showcase to either the customer or the client.

Quantitative environmental targets were absent in all the design briefs. This can be related to the open nature of the design briefs and the high innovation level for the products in this study. Defining quantitative environmental targets in the FE appears to be very difficult for innovation projects where the product parameters are still flexible and the outcome is not well defined.

Another outcome of this study has shown that the design briefing process is not a single activity, but a multi-step process with different actors and decision makers, where different documents, such as the "request for proposal", the "strategic brief", and the real "design brief", jump back and forward between different people and departments in the company. A generic model for the design briefing process (**Figure 6**) was obtained.

The different stages in the design briefing process show different entry points for sustainability; the earlier in the process, the more effective. As product parameters are then still flexible, there is more room for environmental improvement. Having a good view on the decision-making process and who is making the decisions is crucial for integrating sustainability in the FE. They determine what will be written down in the design brief, and what the external/internal team will do in the operational stage of the innovation process. This study has also shown the crucial role of the CEO and the management team in the uptake of environmental sustainability design brief. Without their engagement, nothing sustainable will happen in the design brief. A sustainable product has to be embedded in a strategic sustainable framework, set up in the FE and has to be translated in a design brief. These decisions cannot be taken down the chain. Similar findings are found in the literature [17, 29, 33].

The decision makers need to have a good understanding of the business opportunities, needs, risks, and costs of sustainable product innovation in the FE. Pushing this information upstream in the briefing process can result in a higher success rate on integrating sustainability into the design brief. As long as these topics are not obvious, it will be difficult to convince the stakeholders, and to include it in the design brief. It sounds obvious, but in daily practice, it is rarely the case, due to the characteristics of the FE as explained in 2.1. One strategy to deal with this is "front-loading"; an approach that aims to boost development performance by moving the identification and solving of problems to the first stages of a product development process [34].

As earlier research by [17] showed, the sustainability of the final product cannot be deducted from the design brief. However, the integration of ecodesign targets in the design brief is recommended. With little or no ecodesign components specified in the design brief, it is very hard for the external/internal design and engineering team to take environmental considerations into account during the operational stage of the innovation process. As there is no commitment in the design brief towards the sustainability of the final product from the client, no time, budgets and staff will be allocated on this subject.

The explorative nature of the research in this study has a few limitations. The design brief sample is limited to Belgian Companies, as such for the participants in the focus group and the interview, although the provided insights may also be relevant to other countries.

Acknowledgements

The support for this research from the Ghent University in Belgium and the Flemish Agency for Innovation by Science and Technology is gratefully acknowledged. The author is also indebted to the companies participating in this study and the Design for Sustainability research group at Delft University of Technology for their co-operation and critical feedback. Special thanks to Professor Renee Wever, from the Linköpings Universitet in Sweden, for always believing in me.

Appendix

A. Focus group participants

NACE stands for "Nomenclature Generale des Activites Economiques dans l'Union Europeenne" (General Name for Economic Activities in the European Union) and is the European standard for industry classifications.

Company ID	Sector	NACE code	Category	Title
A	Machine Construction	28940	Multinational	Mechanical Project Manager
A	Machine Construction	28940	Multinational	Project Manager Electrical Engineering
B	Distribution	46499	SME	Designer
C	Industrial Automation	71121	Multinational	Business Manager
D	Construction	711	SME	Engineering and Planning
E	Metal Processing	29320	Multinational	Product Group Manager
F	Plastics Industry	20160	SME	Mechanical Engineering
G	Consumer Electronics	26400	Multinational	Chief Technology Officer
H	Plastics Industry	20160	Multinational	Quality Engineer
I	Electronic Components	26110	Multinational	Design Expert
J	Metal Processing	31091	Multinational	Mechanical and industrial engineering (Owner)
K	Furniture	46150	SME	Technical Sales Employee
L	Building Materials	46731	SME	Head of product management
M	Building Materials	2332	SME	Senior Product Manager
N	Industrial Design	741	SME	Designer

B. Interview guide: topic list

Background interviewee

- Job title, job content, seniority, and educational background

Company profile

- Employees on the staff list
- Active in which areas
- Clients
- Product portfolio
- Specialization of the company
- Number of offices and location

FE innovation process

- General description FEI
- FE activities
- Duration of the different phases
- Role of the respondent in the process
- FE tools
- Experiences in frontloading

Design briefing process

- Starting point/ end point/ steps in between
- Filter mechanisms
- Decision-making and responsibility
- Characteristics of the design brief
- Entry points for sustainability

Vision and strategy on design for sustainability

- Ambition
- Implementation approach in the FE
- Internal/external drivers
- Factors of resistance
- Used tools
- Knowledge management

Author details

Kristel Dewulf
Department of Industrial Systems Engineering and Product Design, Ghent University, Ghent, Belgium

*Address all correspondence to: kristel.dewulf@ugent.be

References

[1] Reid SE, De Brentani U. The fuzzy front end of new product development for discontinuous innovations: A theoretical model. Journal of Product Innovation Managment. 2004;**21, 3**(3):170-184

[2] Koen P, Ajamian G, Burkart R, Clamen A, Davidson J, D'Amore R, et al. Providing clarity and a common language to the "fuzzy front end". Research-Technology Management. 2001;**44**(2):46-55

[3] Phillips PL. Creating the Perfect Design Brief: How to Manage Design for Strategic Advantage. New York, USA: Allworth Design Management Institute; 2004

[4] Hassi L, Peck D, Dewulf K, Wever R. 'Sustainable Innovation, Organization and Goal Finding', Procedings of Joint Actions on Climate Change. Denmark: Aalborg; 2009

[5] Kim J, Wilemon D. Focusing the fuzzy front-end in new product development. R&D Management. 2002;**32**(4):269-279

[6] Crawford CM, Di Benedetto CA. New Products Management. 8th ed. Irwin: McGraw-Hill; 2005

[7] Jacoby A. 'Performance in the Front-End of Innovation: Linking Strategy to Requirements', Doctoral Dissertation. Antwerp: University of Antwerp; 2012

[8] Herstatt C, Verworn B. The fuzzy front end of innovation. In: Bringing Technology and Innovation into the Boardroom. London: Palgrave Macmillan; 2004

[9] von Hippel E. Wettbewerbsfactor Zeit; Moderne Industrie. In: Herstatt C, Verworn B, editors. (2001). The Fuzzy Frond End of Innovation. Hamburg, Germany: Arbeitspapier Nr. 4, Technische Universität Hamburg; 1993

[10] Koen P, Bertels H. Front End of Innovation. Wiley-Blackwell: Wiley International Encyclopedia of Marketing; 2010

[11] Khurana A, Rosenthal S. Towards holistic "front ends" In new product development. The Journal of Product Innovation Management. 1998;**15**:57-74

[12] Verworn B. A structural equation model of the impact of the "fuzzy front end" on the success of new product development. Research Policy. 2009;**38**(10):1571-1581

[13] Wormald W. Front End Industrial Design—Developing New Tools and Models for Industrial Designers to Operatie at the Front End of New Product Development. Singapore: National University of Singapore; 2010

[14] Zarney K. The Core creative concept in branding: A streamlined approach. Design Management Journal. 2002;**13**(4):39

[15] Parkman I. Two Essays Examining Design Briefs as Knowledge-Based Assets: Content and Cross-Functional Collaboratio. Oregon, USA: University of Oregon; 2010

[16] Blyth A, Worthington J. Managing the Brief for Better Design. 2nd ed. London, United Kingdom: Routledge; 2010

[17] Petala E, Wever R, Dutilh C, Brezet H. The role of new product development briefs in implementing sustainability: A case study. Journal of Engineering and Technology Management. 2010;**27**:172-182

[18] Bocken NMP, Farracho M, Bosworth R, Kemp R. The front-end of eco-innovation for eco-innovative small and medium sized companies.

Journal of Engineering and Technology Management. 2014;**31**(0):43-57

[19] Dewulf K. In: Coelho D, editor. Sustainable Product Innovation: The Importance of the Front- End Stage in the Innovation Process. Advances in Industrial Design Engineering, Rijeka: InTech; 2013

[20] Dewulf K, Wever R, Brezet H. Greening the design brief. In: Design for Innovative Value Towards a Sustainable Society, Proceedings of EcoDesign 2011: 7th International Symposium on Environmentally Conscious Design and Inverse Manufacturing. The Netherlands: Springer; 2012. pp. 457-462

[21] Hallstedt S, Ny H, Robèrt K, Broman G. An approach to assessing sustainability integration in strategic decision Systems for Product Development. Journal of Cleaner Production. 2010;**18**(8):703-712

[22] Hallstedt SI, Thompson AW, Lindahl P. Key elements for implementing a strategic sustainability perspective in the product innovation process. Journal of Cleaner Production. 2013;**51**(0):277-288

[23] Hassi L, Wever R. Practices of a 'Green' front end. A gateway to environmental innovation. In: Proceedings of ERSCP-EMSU 2010. Delft, The Netherlands: Delft University of Technology; 2010

[24] Lofthouse V. Investigation into the role of Core industrial designers in ecodesign projects. Design Studies. 2004;**25**(2):215-227

[25] O'Hare J. Eco-Innovation Tools for the Early Stages: An Industry-Based Investigation of Tool Customisation and Introduction. Bath, United Kingdom: University of Bath; 2010

[26] van Bommel HWM. A conceptual framework for Analyzing sustainability strategies in industrial supply networks from an innovation perspective. Journal of Cleaner Production. 2011;**19**(8):895-904

[27] Wever R, Casper B. Design for sustainability in the fuzzy front end. In: Proceedings of Sustainable Innovation 2007. The Center for Sustainable Design. United Kingdom: Surrey; 2007. pp. 199-205

[28] Storacker A, Wever R, Dewulf K, Blankenburg D. Sustainability in the front-end of innovation at design agencies. In: Proceedings of Ecodesign 2013, 8th International Symposium on Environmentally Conscious Design and Inverse Manufacturing, Jeju Island, Korea; 2013.

[29] Dewulf K. Play it forward. A game-based tool for sustainable product and business model innovation in the fuzzy front end. In: Proceedings of ERSCP 2010, Delft, The Netherlands; 2010.

[30] van Hemel C, Cramer J. Barriers and stimuli for Ecodesign in Smes. Journal of Cleaner Production. 2002;**10**(5):439-453

[31] Sleeswijk Visser F, Stappers PJ, van Der Lugt R, Sanders E. Contextmapping: Experiences from practice. CoDesign. 2005; **1**(2):119-179

[32] British Design Council. A Study of the Design Process. London, UK: Design Council; 2006

[33] Driessen PH. Green product innovation strategy. In: CentER Dissertation Series. The Netherlands: Tilburg University; 2005. p. 150

[34] Thomke S, Fujimoto T. The effect of front-loading problem solving on product development performance. Journal of Product Innnovation Management. 2000;**17**(2):128-142

Section 2

Additive Manufacturing

Chapter 5

Prologue: The New Era of Sintering

Eleonora Santecchia and Mohsen Mhadhbi

1. Introduction

Nowadays, the new era of development of the sintering process has begun. The sintering of metal and ceramic powders plays an important role in our industry.

Sintering is a one-way processing technique which allows the production of density-controlled materials and components from metal or/and ceramic powders through particle bonding. This process of powder consolidation is quite famous and has been applied in different fields for more than 5000 years [1–4].

In general, sintering processes can be classified according to the material state during the sintering process as solid-state sintering and liquid phase sintering. The ability to achieve consolidation without melting is made possible by the thermal activation of mass transport processes driven by reduction of surface and grain boundary energies. To optimize thermal activation and attain high density with concomitant strength, sintering is carried out at high temperatures, relative to the melting point of the material. Optimal parameters for sintering result from a combination of material variables and process parameters [5], as shown in **Table 1**.

The sintering process generates an increase in the compact strength and rigidity, together with a dimensional change of the powder particles, with the formation of necks between contacting spheres. Typically, longer sintering times produce larger necks and, therefore, samples with higher strength.

Money-saving reasons moved the scientific research toward a variety of means of thermal activation and, in particular, electric current, in order to produce high-density objects in a faster way and using lower temperatures. Spark plasma sintering (SPS) and pulsed electric current sintering (PECS) techniques are only a few examples of how far the research is pushing the performance of sintering, at both fundamental and applied levels [7–9].

In terms of heating methods, the most common can be synthesized as follows: (i) contact methods, such as thermal conduction, radiation, or convection and (ii) noncontact heating methods, such as induction, radio frequency (RF), or microwave heating.

This book consists of six sections; the new era of sintering is presented at the beginning of the book. This first section opens with the introduction on sintering followed by an overview on the different sintering techniques and thermodynamics and kinetics of sintering. The second section focuses on the influence of sintering on microstructure and mechanical properties. The third section provides a comprehensive summary on the solid-state sintering of materials (intermetallics, ceramics, metals, and composites). The fourth section is dedicated to sintering-based 3D printing as a new technology. The fifth section addresses on the composite sintering and applications. The mathematical models and numerical methods for continuous sintering approaches are fully discussed in the last section.

Material (powder) parameters		Process parameters
Physics	**Chemistry**	
Shape	Composition	Temperature
Size	Impurity	Time
Size distribution	Stoichiometry	Pressure
Agglomeration	Homogeneity	Atmosphere
Mixedness		Heating/cooling rate

Table 1.
Summary of the major sintering parameters (adapted from [6]).

The book aims to provide informative chapters to the readers, researchers, and material engineering and industrial material scientists. The chapters in this book are from specialists in their respective disciplines. In addition, this book is very important for the diffusion of the scientific knowledge.

2. Sintering techniques

Sintering is the process of consolidating powder compact by a thermal treatment to obtain materials with special properties. It is one of the widely used techniques in the powder metallurgy and ceramic processing. These techniques enable heating through interaction between electromagnetic field and materials.

2.1 Conventional sintering

Powder forming and, in particular, pressing are among the best ways to build flat objects. At first, the metal powder is pressed into a die, having a geometry close to the one of the final part, together with binders (to increase compact ability) at room temperature to form the so-called green part, which is typically strong enough to be handled gently. The sintering of the green part can be divided into three stages: (i) preheating; (ii) sintering, at a temperature maintained for a time depending on the strength of the bond which needs to be obtained; and (iii) cooling. Usually all the three stages are performed in a controlled atmosphere furnace, in order to prevent phenomena such as oxidation or unwanted chemical reactions. Sintering of stainless steel and refractories is usually performed under vacuum conditions.

As the name might suggest, in liquid sintering a portion of the material is in the liquid phase. This procedure is typically used for cermet, metal, and ceramic sintering. During liquid phase sintering, a liquid phase coexists with a particulate solid at sintering temperature. The goodness of sintering in this case depends strongly on the wetting properties of the liquid part [10].

2.2 Electric current-aided sintering

Heating induced by electric current has some interesting advantages with respect to conventional heating sources such as the lower sintering temperature, which allows to process nanometric powders, shorter time duration, and better material properties [11–13].

Pulsed electric current sintering (PECS) is characterized by the simultaneous action of a current-induced heating and a uniaxial pressure. PECS allows to reach high heating rates and influences mass transport. Typically, a pulsed DC current is applied with a relatively low voltage (below 10 V), and the current is applied by pulsing patterns. The pulsing pattern is made up of a sequence of pulses (3.3 ms each) followed by an interruption of current, i.e., a pulse pattern of 12-2 means that 12 pulses are applied, followed by a duration of 2 pulses where the current is not applied [9].

The same simultaneous application of heating and uniaxial pressure is performed during spark plasma sintering (SPS) high densification at temperatures which are typically around 200°C lower than in conventional sintering. Thus, SPS is a new powder consolidation technique used to fabricate bulk shapes and nanostructured materials. This method uses uniaxial pressure and pulsed direct electric current to consolidate the powders at short sintering time and a relatively low sintering temperature compared to conventional hot pressing sintering techniques. The short sintering time is suitable for preserving the amorphous structure without undesirable phase transformations. During SPS heating is due to a pulsed DC current passing through the die which contains the powder, while pressure is applied on it. The characteristics, therefore, include the high heating rate, the application of a pressure, and the effect of the current. While similar for some features to conventional sintering, the SPS process is typically characterized by a higher heating rate such as 300°C/min compared to a maximum of 10°C/min reachable during conventional sintering. Therefore, high relative densities can be obtained in a very short time, allowing to sinter nanometric powders, nanostructured ceramics, or nanocomposites, avoiding considerable grain growth [8].

2.3 Microwave sintering

Microwave sintering, which includes heating and sintering, is a powerful process for sintering ceramic, ceramic composites, cermets, and metals. In microwave sintering the heating is obtained through a noncontact method which results in enhancement of the process in terms of reaction and diffusion kinetics, shorter cycle time, finer microstructures, and other unique features leading to considerable improvement in the mechanical properties and further energy savings [14–16]. While developed for ceramic, inorganic, and polymeric materials, microwave sintering is now employed also for all metal powders [17–21].

During microwave sintering, heating takes place via absorption/coupling.

of the microwave field followed by the so-called volumetric heating (viz., heating of the material as a whole) due to the conversion of the electromagnetic energy into.

thermal energy. In this particular case, heating is generated within the material in an instantaneous way, which depends strongly on the material properties. Therefore, with respect to conventional sintering, the heating profile is inside-out instead of outside-in [22].

2.4 Hot isostatic pressing

Hot isostatic pressing (HIP) is a manufacturing technique based on the simultaneous application of temperature and pressure to materials (usually powders) for a definite time to increase the density of materials. This process was invented in 1955 to improve aircraft systems and nuclear industry. Today it became an emerging technology in the processing of high-density powders that is used in aerospace, automotive, medical defense, etc. [23]. This technique has several advantages and involves highly complex shapes of finished parts, the powders are consolidated at lower temperatures achieving higher densities, the finished parts have homogenous density, the high gas density results in rapid heating and shorter time, and the brittle materials can be processed because of the more uniform heating [24]. Thus, HIP involves the simultaneous application of high temperature and pressure cycles [25]. In the first cycle (cold loading cycle), the temperature is increased some time after the pressure reaching their peak at the same time to give good geometric control in sheet metal encapsulation. In the second cycle (hot loading cycle), the pressure is applied after the temperature has reached its desired value. In the third cycle, the temperature is raised only after pressure reaches its desired value. In the final cycle, the pressure and the temperature are increased simultaneously to reduce the processing time.

The advantage of HIP is the reduction of production times and the variation of properties in the solid, obtaining almost finished parts and parts with complex shapes or small sizes due to its isostatic processing. It is noteworthy in this instance that the model used will be validated over similar powder forming processes where the application of temperature and pressure is performed simultaneously [26–28].

2.5 Sintering-based 3D printing

3D printing is a new technology, which appeared since the 1980s, used to produce bulk materials (metals, ceramics, composites, etc.) with complex geometry. This technique is based on extrusion followed by sintering. Thus, the advantages of this method are the low cost and the versatile printing strategy. 3D printing allows to fabricate parts in a discrete layer by layer or line by line or point by point from 3D computer-aided design (CAD) models. It is the unique technique that enables to fabricate precise and highly complex parts that are difficult to obtain using traditional techniques [29]. Three-dimensional printing technology is expected to solve the limitations that are inevitably encountered when using traditional methods. 3D printing, also known as additive manufacturing, is based on the principle of layered manufacturing, in which materials are overlapped layer by layer [30]. This technology can be used to quickly fabricate components with any complex shape by accurately accumulating material using solid modeling according to a CAD model or computed tomography (CT) scan under computer control [31]. The 3D printing industry has recently exploded due to the reduced manufacturing costs of 3D printers and to their improved printing precision and speed, allowing for huge advances in medical equipment, implant material, and cell printing.

3. Thermodynamics and kinetics of sintering

The basic of sintering is the formation of necks between powder particles. While neck formation depends on system thermodynamics, the rate of sintering is mostly due to the temperature of the process. At room temperature, the atoms in a material are not noticeably mobile, so the particles do not sinter. When a material is heated at a temperature close to its melting point, the atoms increase their mobility and produce bonding formation, which lowers the overall system energy. The energy changes in sintering are small so that the rate of change during sintering is slow [22].

At the beginning of sintering, the interparticle neck grows to the point where its size is less than one-third of the particle size, often associated with a little dimensional change. Sintering proceeds with necks growing to a dimension which is larger than one-third of the particle size but still lower than half of the particle dimension, corresponding to a density between 70 and 92% for spheres. The pores are tubular and connected to the external surface, and the green sintering body is compact but not fully dense so that gas can still pass through it. In the end, the final stage of sintering corresponds to the elimination of surface pores, while internal isolated pores are filled with process atmosphere.

Sintering is mainly driven by the reduction of surface energy associated with the decrease in the pore-solid surface area. Locally, there is a chemical potential shift on each component of surface on the interface, from the value for a flat interface in the same system, which is proportional to the local mean curvature H at the surface element, as shown by thermodynamics.

Assuming that local equilibrium exists at each interfacial element, the chemical potential of vacancies in a volume close to a curved surface element is given by Eq. (1):

$$\mu_v(H) = \mu_v^0 + 2\gamma VH \tag{1}$$

where V is the molar volume of the solid phase, γ is the specific interfacial free energy, and $\mu_v{}^0$ is the chemical potential of vacancies adjacent to a flat surface at a given temperature and external pressure of the system. The chemical potential shift reflects in a change in the molar concentration of vacancies ($c_v(H)$), which form a dilute solution. Given that a system with a flat surface has a local mean curvature equal to zero, the shift in vacancy molar concentration can be written as Eq. (2):

$$c_v(H) = c_v^0 - \frac{2\gamma V c_v^0}{kT} H \tag{2}$$

where T is the absolute temperature and k is Boltzmann's constant. This theoretical treatment shows that the local mean curvature on the surface determines the distribution of vacancy concentrations in volume elements close to the pore-solid interface. Therefore, for convex surface elements, the local stress state is compressive, and $c_v(H) < c_v{}^0$, while for concave surface elements, the local stress state is the opposite, and $c_v(H) > c_v{}^0$. The vacancy molar concentration distribution modifies their flow distribution, and grain boundaries act like sinks, while surface elements defined by $H = 0$ are source of vacancies. The densification in conventional sintering is, therefore, due to the vacancy annihilation at the grain boundaries.

In terms of kinetics, there is a variety of mechanisms which control the sintering process at different levels, such as volume, grain boundary and surface diffusions, plastic deformation, and vapor transport. Densification is mainly influenced only by the first two mechanisms, being defined as a decrease in volume of the pore-solid structure. Plastic deformation generates only a limited amount of surface tensions and gives, namely, a negligible contribution to sintering densification control. Vapor transport is the dominant mechanism when the solid phase has a high vapor pressure, although it is not able to produce densification since both vacancy sources and sinks are on the surface. This limitation is verified also for the surface diffusion mechanism [22].

4. Conclusions

Conventional and new sintering systems, based on various mechanisms of heat generation, are changing the world of powder metallurgy. The combination of different parameters, which can be related to the materials used or to the particular process, allows the production of objects having properties which can make them suitable for different applications. This book inspires materials scientists to carry on efforts concerning sintering.

Acknowledgements

We would like to thank the Author Service Manager Ms. Marina Dusevic for his assistance in communicating. We also thank our institutions for give us support.

Conflict of interest

All authors have seen and approved the manuscript being submitted. We warrant that the chapter is the authors' original work. We warrant that the chapter has not received prior publication and is not under consideration for publication elsewhere. On behalf of all the co-authors, the corresponding author shall bear full responsibility for the submission.

This work has not been submitted for publication, nor has it been published in whole or in part elsewhere. We attest to the fact that all authors have contributed significantly to the work, have read the manuscript, attest to the validity and legitimacy of the data and its interpretation, and agree to its submission to the IntechOpen.

All authors agree that author list is correct in its content and order and that no modification to the author list can be made without the formal approval of the editor.

Thanks

We thank IntechOpen Publisher for accepting this work to participate in the book project under the working title "Design Engineering and Manufacturing", ISBN 978-1-78985-866-2.

Author details

Eleonora Santecchia[1] and Mohsen Mhadhbi[2*]

1 Department of Materials, Environment Sciences and Urban Planning SIMAU, Marche Polytechnic University, Ancona, Italy

2 Laboratory of Useful Materials, National Institute of Research and Physical-Chemical Analysis, Ariana, Tunisia

*Address all correspondence to: mhadhbi_mohsen@yahoo.fr

References

[1] Upadhyaya GS. Sintering Fundamentals, Switzerland. Trans Tech Publications; 2009

[2] Pero-Sanz Alroz JA, Verdeja-Gonzalez JY, Sanch-Martinez JP, Vilela N. Melting and sintering platinum in the 18th century: The secret of the Spanish. Journal of Metals. 1999;**51**(10):9-13

[3] Liu B, Li J, Ivanov M, Liu W, Ge L, Xie T, et al. Diffusion-controlled solid-state reactive sintering of Nd:YAG transparent ceramics. Ceramics International. 2015;**41**:11293-11300

[4] Pötschke J, Säuberlich T, Vornberger A, Meese-Marktscheffel JA. Solid state sintered nanoscaled hard metals and their properties. International Journal of Refractory Metals & Hard Materials. 2018;**72**:45-50

[5] Gültekin EE. The effect of heating rate and sintering temperature on the elastic modulus of porcelain tiles. Ultrasonics. 2018;**83**:120-125

[6] Kang S-JL. Sintering—Densification, Grain Growth, and Microstructure. UK: Elsevier; 2005

[7] Garay JE. Current-activated, pressure-assisted densification of materials. Annual Review of Materials Research. 2010;**40**:45-68

[8] Munir ZA, Anselmi-Tamburini U, Ohyanagi M. The effect of electric field and pressure on the synthesis and consolidation of materials: A review of the spark plasma sintering method. Journal of Materials Science. 2006;**41**:763-777

[9] Munir ZA, Quach DV. Electric current activation of sintering: A review of the pulsed electric current sintering process. Journal of American Ceramic Society. 2011;**94**(1):1-19

[10] German RM. Sintering Theory and Practice. New York: John Wiley & Sons; 1996

[11] Anselmi-Tamburini U, Garay JE, Munir ZA, Taca A, Maglia F, Spinolo G. Spark plasma sintering and characterization of bulk nanostructured fully stabilized zirconia: Part I. Densification studies. Journal of Materials Research. 2004;**19**:3255-3262

[12] Uskoković DP, Palmour H III, Spriggs RM. Science of Sintering: New Directions for Materials Processing and Microstructural Control. New York: Springer; 1989

[13] Ghasali E, Sangpour P, Jam A, Rajaei H, Shirvanimoghaddam K, Ebadzadeh T. Microwave and spark plasma sintering of carbon nanotube and graphene reinforced aluminum matrix composite. Archives of Civil and Mechanical Engineering. 2018;**18**:1042-1054

[14] Clark D, Sutton WH. Microwave processing of materials. Annual Reviewof Materials Science. 1996;**26**:299-331

[15] Schiffman RF. Commercializing microwave systems: Paths to success or failure. Ceramic Transactions. 1995;**59**:7-17

[16] Agrawal DK. Microwave processing of ceramics: A review. Current Opinions in Solid State Materials Science. 1998;**3**(5):480-486

[17] Roy R, Agrawal D, Cheng J, Gedevanishvilli S. Full sintering ofpowdered-metal bodies in a microwave field. Nature. 1999;**399**:668-670

[18] Anklekar RM, Agrawal DK, Roy R. Microwave sintering and mechanical

properties of P/M steel. Powder Metallurgy. 2001;**44**(4):355-362

[19] Sethi G, Upadhyaya A, Agrawal D. Microwave and conventional sintering of pre-mixed and prealloyed Cu-12Sn bronze. Science of Sintering. 2003;**35**(49):49-65

[20] Takayama S, Saiton Y, Sato M, Nagasaka T, Muroga T, Ninomiya Y. Microwave sintering for metal powders in the air by non-thermal effect. In: 9th International Conference on Microwave and High Freq Heating. 2003. pp. 369-372

[21] Anklekar RM, Bauer K, Agrawal DK, Roy R. Improved mechanical properties and microstructural development of microwave sintered copper and nickel steel PM parts. Powder Metallurgy. 2005;**48**(1):39-46

[22] Fang ZZ. Sintering of Advanced Materials: Fundamentals and Processes. UK: Woodhead Publishing; 2010

[23] Atkinson HV, Davies S. Fundamental aspects of hot isostatic pressing: An overview. Metallurgical and Materials Transactions A. 2000;**31**:2981-3000

[24] Loh NL, Sia KY. An overview of hot isostatic pressing. Journal of Materials Processing Technology. 1992;**30**:45-65

[25] Buekenhout L, Alt P. Hot isostatic pressing of metal powders. Key Engineering Materials. 1989;**29-31**:207-224

[26] Srivatsan T, Ravi B, Naruka A, Riester L, Petraroli M, Sudarshan T. The microstructure and hardness of molybdenum powders consolidated by plasma pressure compaction. Powder Technology. 2001;**114**(1-3):136-144

[27] Ohser-Wiedemann R, Martin U, Seifert HJ, Müller A. Densification behaviour of pure molybdenum powder by spark plasma sintering. International Journal of Refractory Metals and Hard Materials. 2010;**28**(4):550-557

[28] Mouawad B, Soueidan M, Fabrègue D, Buttay C, Bley V, Allard B, et al. Full densification of molybdenum powders using spark plasma sintering. Metallurgical and Materials Transactions A. 2012;**43**(9):3402-3409

[29] Gibson I, Rosen D, Stucker B. Additive Manufacturing Technologies: 3D Printing, Rapid Prototyping, and Direct Digital Manufacturing. 2nd ed. New York: Springer; 2014. 498 p

[30] Qian Y, Hanhua D, Jin S, Jianhua H, Bo S, Qingsong W, et al. A review of 3D printing technology for medical applications. Engineering. 2018;**4**:729-742

[31] Yeong WY, Chua CK, Leong KF, Chandrasekaran M, Lee MW. Indirect fabrication of collagen scaffold based on inkjet printing technique. Rapid Prototyping Journal. 2006;**12**:229-237

Chapter 6

Utilization of Additive Manufacturing to Produce Tools

Kıvılcım Ersoy and Berk Barış Çelik

Abstract

In the last decade, customized design and small series production gained importance in various industries. The production of these special tools becomes one of the most important costs for the production process. With the advances in additive manufacturing (AM) technologies, tools can be produced efficiently in short lead times and costs with additive manufacturing. In this chapter, first, an overview on the additive technologies to produces tools, also called rapid tooling, will be given. The advantages as well as disadvantages will be discussed. Following that, on an example of metal forming tools, different materials coupled with different additive production techniques will be compared. Also, most important points will be highlighted to select the most appropriate tool material and manufacturing method. Finally, a methodology to identify the tool life will be suggested, and its validation and verification on a simplified deep drawing geometry will be depicted. The comparison of numerical prediction and experimental results are shown to be in good agreement.

Keywords: rapid tooling, tool design, additive manufacturing, sheet metal forming tools, tool life prediction

1. Introduction

In this section, first of all the history and the literature data of the rapid tooling will be mentioned. Then, the AM using methods of rapid tooling will be explained in detail. Finally, examples to be taken into consideration when making the tool will be given basic flow chart of the processes can be seen in **Figure 1**.

AM techniques for tool production (known in literature rapid tooling [RT]) provide efficiency in terms of time and cost, while RT is often the best known manufacturing method for complicated structures in low numbers [2]. AM techniques for manufacturing, prototyping, and tool production methods can also allow producers to manufacture high-quality elements in a short period [3, 4]. In today's world, with the technology advancing rapidly, companies are looking for ways to manufacture varied and complicated items with high quality while reducing the cost and time needed [5].

AM techniques for tool production defines a method resulting from mixing AM techniques for prototyping with standard tooling disciplines to produce a die rapidly or components of a functional model from CAD information at a reduced cost and in less time than conventional machining methods.

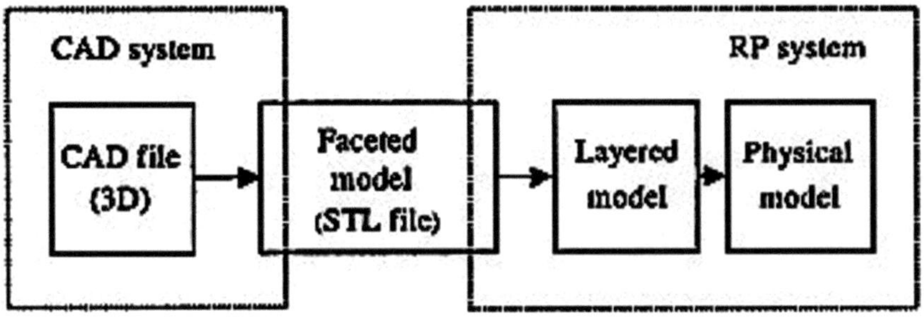

Figure 1.
Data transfer between the computer-aided design (CAD) and the RP system [1].

2. Additive methods to produce tools (rapid tooling)

There are two main categories. One category includes indirect methods using AM techniques for prototyping master designs to manufacture a mold, while the other category is a direct strategy, where the rapid prototyping machine constructs the real inserts of the core and cavity mold.

Due to its market potential, many businesses prefer AM techniques for tool produce and growth. Each method comes with a number of strengths that are contracted by limitations. Yet these advances cause a flurry of requests from businesses in the Americas, Europe, Asia, and other advanced markets due to their potential effect. Meanwhile, in both of these methods, numerous facilities are working hard to determine whether the time is correct to phase further information can be seen in **Table 1**.

2.1 Indirect methods of rapid tooling

To achieve various lead times, expenses, and process capabilities, there occurred several pattern-based methods in manufacturing a mold. To compose a pattern, the precision of RP processes gets considered with each part structure and the precision of the RP process preferred:

- Vacuum casting.
- Kirksite tooling.
- Electroforming.
- RTV silicone rubber molds.
- Wax injection molding.
- RIM.
- Spin casting.
- Plaster molds.
- Rapid solidification process.
- Sprayed steel.
- Cast resin tooling.

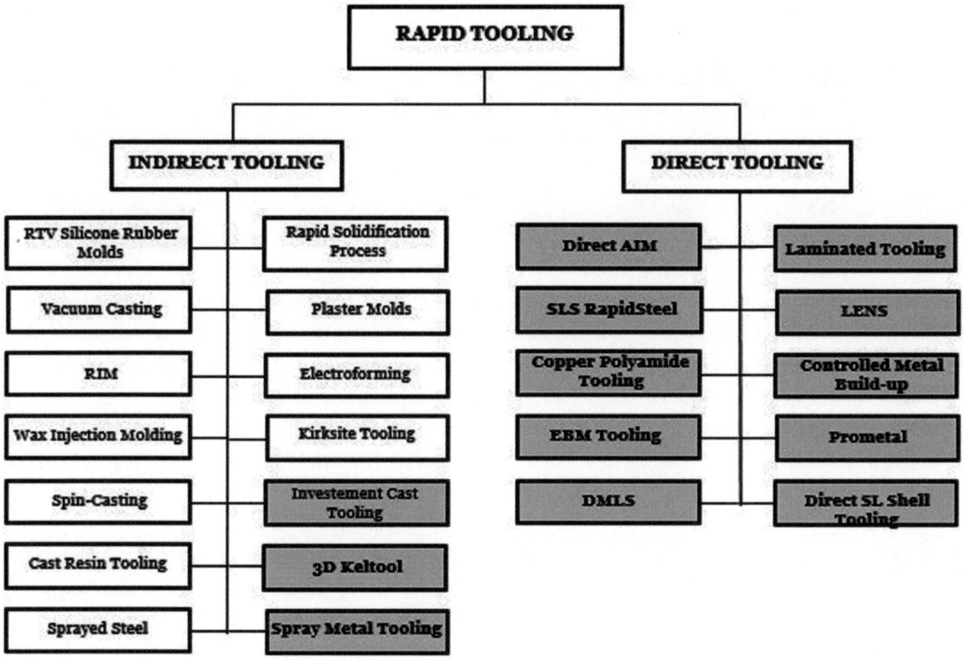

Table 1.
Methods of rapid tooling.

2.1.1 Investment cast tooling

Investment casting has been used for thousands of years as a manufacturing method and is particularly suited to finely detailed metal parts being formed and high melting point metals being used.

Founded in 1947, Vaupell uses SLA patterns for casting to quickly deliver 3D-printed patterns for casting to customers in aerospace industry. Here, AM (SLA) is used to manufacture a sacrificial model. Using this sacrificial model, investment casting is carried out as it would traditionally be the model with a ceramic coating, scorching the model, filling the remaining ceramic shell with molten metal, and removing the ceramic case after the metal has been solidified and cooled [6].

2.1.2 3D Keltool

As a first step in the 3D Keltool process, cavity mold inserts and a core get designed in CAD to get followed by manufacture of the core and cavity patterns with stereolithography or some other processes. Once these core and cavity patterns meet the requirements of the surface, silicone rubber cast gets applied on them to form molds in which a mixture of a metal powder and binder get poured and packed.

The instruments used in this process illustrate very qualified surface finish and definition. In general, lead time results in shorter time period than the conventional tooling. The main limitation is the size restriction. The size of a mold insert could be 100 × 150 × 215 mm (4 × 5.9 × 8.5 inches) at most. If the x and/or y dimension gets smaller, the z-direction has the possibility to extend to 145 mm (5.75 inches). To manufacture larger items, some tool makers have two or more cheek-to-cheek inserts in a die base [6].

2.1.3 Spray metal tooling

Metal spray molds had been used effectively in low-pressure procedures including RIM, vacuum forming, and rotational molding. Currently, the improvements in spray metals and spraying methods give rise to its use in injection molding.

The order of steps is like the one applied to manufacture epoxy molds, with the exception that the pattern is sprayed first with metal and then supported by epoxy resin filled with metal. The process of spraying with, for example, an electrical compressed air gun gets applied until the required shell thickness is met (0.5 mm + – mm is reported). In this technique, the material consisting of pattern must have increased strength and durability to withstand the thermal impact inherent: in this regard, the use of ABS FDM masters, polycarbonate SLS masters, and machinable wax gave positive results some of the indirect methods which uses AM, and their advantages and disadvantages are shown in **Table 2**. In addition to them,

METHOD	ADVANTAGES	DISADVANTAGES
Investment Cast Tooling	Suited to forming finely detailed metal components and working with high-melting point metals.	Size limitations and usually only smaller castings can be made.
3D Keltool	Very good definition and surface finish.	Size limitations.
Spray Metal Tooling	Excellent abrasion resistance (nickel-spray tooling).	Required skilled or semi-skilled labor so the labor cost may be relatively higher than other rapid tooling techniques.

Table 2.
Advantages and disadvantages of indirect methods, which use AM.

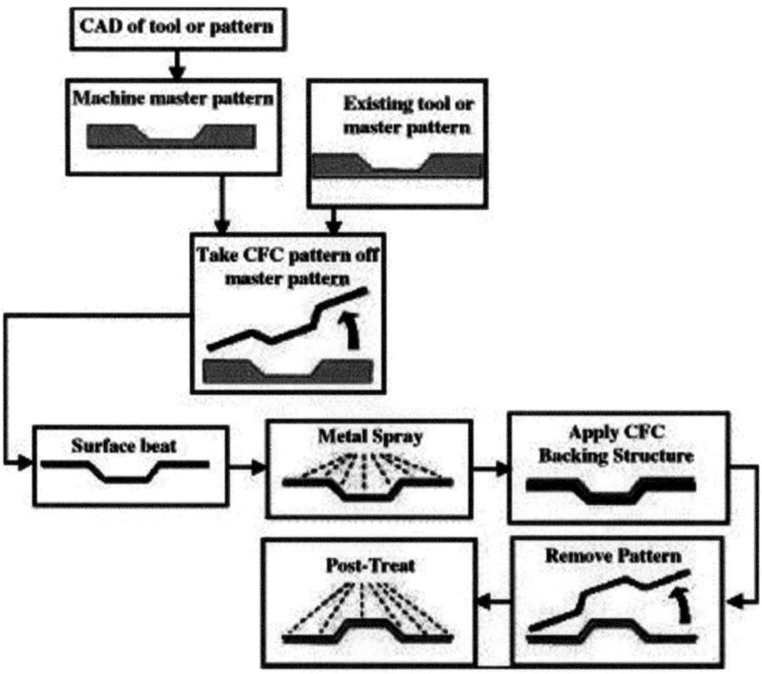

Figure 2.
Spray metal tooling [7].

a reflective coating could be protective. Also, SLA models were tried process of the spray metal tooling depicted in **Figure 2** [6].

2.2 Direct methods of rapid tooling

All the past AM techniques mentioned include indirect production of a master model producing the tool. The time required to manufacture and complete its pattern is one of the major issues with the manufacturing tool. Moreover, such replication methods can still cause increase in inaccuracies. As a result, although the most of direct tooling techniques have constraints, corporations ask directly produced tooling.

Applying additive "layer manufacturing" methods, further properties may get included in the item, which cannot be acquired with conventional tooling techniques. Conformal heating (or cooling) channels provide the system with the ability to heat or cool the exact points needed, and it is the most critical property of it. It is shown that conformal channels could result with the reduction in cycle times of injection molding by up to 40% [6].

2.2.1 Direct AIM

Rather than making a master stereolithography pattern around which a material is cast, it also is possible to build the cavity directly on the stereolithography machine. This method has been described as Direct AIM by 3D Systems (Valencia, CA). (AIM stands for ACES Injection Molding. ACES stands for "accurate clear epoxy solid," which is a stereolithography construction style.) Even though they are not as strong or difficult as standard tools, various thermoplastics can get injected into those cavities so that elements to get used could get manufactured. Even though only less abrasive and lower melting polymers can get molded, studies are going on in order to increase its applicability [6].

2.2.2 SLS RapidSteel

Just as a cavity can be produced directly by stereolithography, the laser sintering method can also be used to construct tool cavities directly. Digital core and cavity geometry models are developed and sent to a Sinterstation manufacturing device in RapidSteel powder with DTM's RapidSteel (also known as RapidTool, earlier similar methods are known as Indirect Metal Selective Laser Sintering). This material comprises of mild stainless-steel particles that are covered with a thin layer of a material for a polymer binder. The Sinterstation generates green components that fire in a furnace afterwards. The furnace removes the polymer binder and by capillary action infiltrates bronze into the inserts of the mold. This method generates a completely thick tool consisting of approximately 60% steel and 40% bronze. Then the inserts are completed, drilled for ejector pins, and fitted to the base of the mold.

The technique generates a durable mold that can be used as well as die-casting apps for injection mold tooling. Hundreds of aluminum, zinc, and magnesium components were casted using RapidSteel molds. The technique enables complicated geometries, and molds from RapidSteel can resist injection molding circumstances. RapidSteel, however, needs finishing and polishing that can take time scheme of the SLS process can be seen in **Figure 3** [6].

2.2.3 Copper polyamide tooling

The DTM (Austin, Texas) copper polyamide tooling method is consisted of selective laser sintering of a matrix of copper and polyamide powder to manufacture

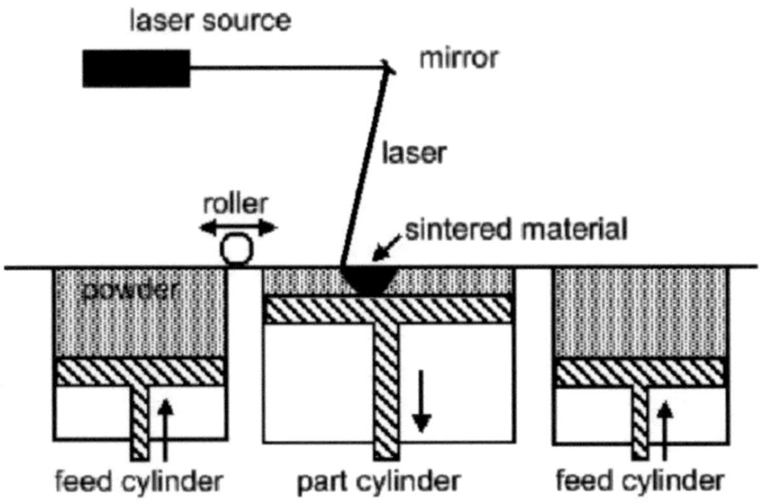

Figure 3.
DTM SLS Sinterstation 2500 plus [8].

a tool. All sintering process happens among the polyamide powder particles. This technique provides improvements in toughness of items unlike some of the other smooth tooling techniques and heat transfer. The copper is not only appropriate for these properties but also can offer some advantages to users like operating a device with pressure and temperature settings which are closer to the environment of the process. The main inconvenience of it is the material's low resistance [7].

2.2.4 EBM tooling

Arcam provides technology to manufacture completely solid metal elements with electron beam melting (EBM). Parts layer by layer are made by the EBM technology using strong electron beam (4 kW power), which melts metal powder. The use of EBM method performed in a vacuum provides users with stress relaxed components with better mechanical, chemical, and material properties the casting and forming.

The method depends on the use of high-level energy thanks to its ability to provide high fusion ability and high productivity. The EBM method is mainly developed in order to process refractory as well as resistant materials (tantalum, niobium, molybdenum, tungsten, vanadium, hafnium, zirconium, titanium) and alloys thereof. It is defined primarily by not only high-speed manufacturing but also complicated geometries of elements with comparable mechanical characteristics to heat-treated products [7].

2.2.5 Direct metal laser sintering

Direct metal laser sintering (DMLS) from EOS consists of metal powders processed directly in a laser sintering machine. The machine manufactures not only tool inserts but also metal parts. Two materials are available for DMSL method and this method depicted in **Figure 4**:

1. Bronze-based materials are preferable for injection molding of up to 1000 elements in various products.

2. Steel-based material which is advantageous for injection molded components of up to 100,000 plastics [7].

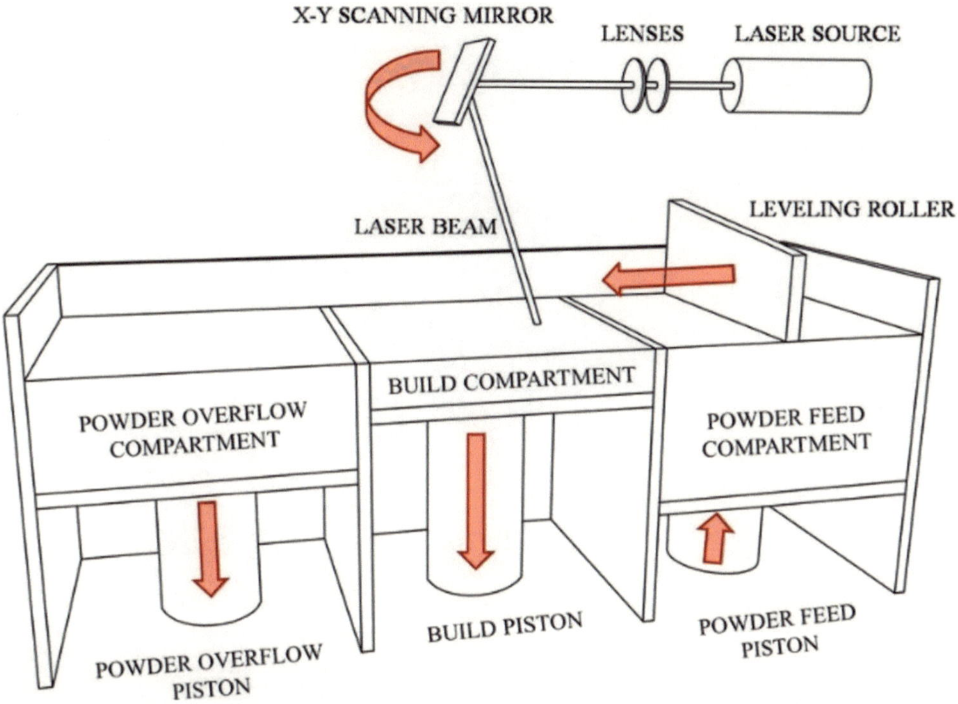

Figure 4.
DMLS 3D printing process [9].

2.2.6 Laminated tooling

Laminated tooling is another applicable option to construct cavities on an AM used prototyping machine. By using a CAD model, sheet layers of metal are sliced, which uses comparable principles to the laminated object manufacturing (LOM) method to multiply slices. Either water-jet or laser cutting techniques are usually applied to obtain the profiles. Manufacturing the molding tool requires CAD model to take the form of the necessary cavity. A mass of laminates can be accomplished to replicate by cutting all the cavity slices into sheet metal. To eliminate complicated post-process cutter path planning, a pseudo-solid cavity in hardened tool steel is manufactured by using either clamping or diffusion bonding related picture can be seen in **Figure 5** [6].

2.2.7 Lens

The Optomec (Albuquerque, New Mexico) laser-engineered net shaping (LENS) system—initially created at the Sandia National Laboratories—is used to build elements into a laser, primarily laser cladding using a metal powder feed. Through a highly intense laser beam into a molten metal pool, a metal powder is injected in this method. The manufacturing method takes place for oxygen-free operation in a low-pressure argon chamber. A movement scheme drives a platform through x and y planes (two-dimensional) as the laser beam traces the cross section of the fabricated portion which can be seen in **Figure 6** [7].

2.2.8 Controlled metal buildup (CMB)

Albrecht Röders GmbH & Co. KG (Soltau, Germany) has marketed a method called controlled metal buildup (CMB). At the Fraunhofer Institute for Production

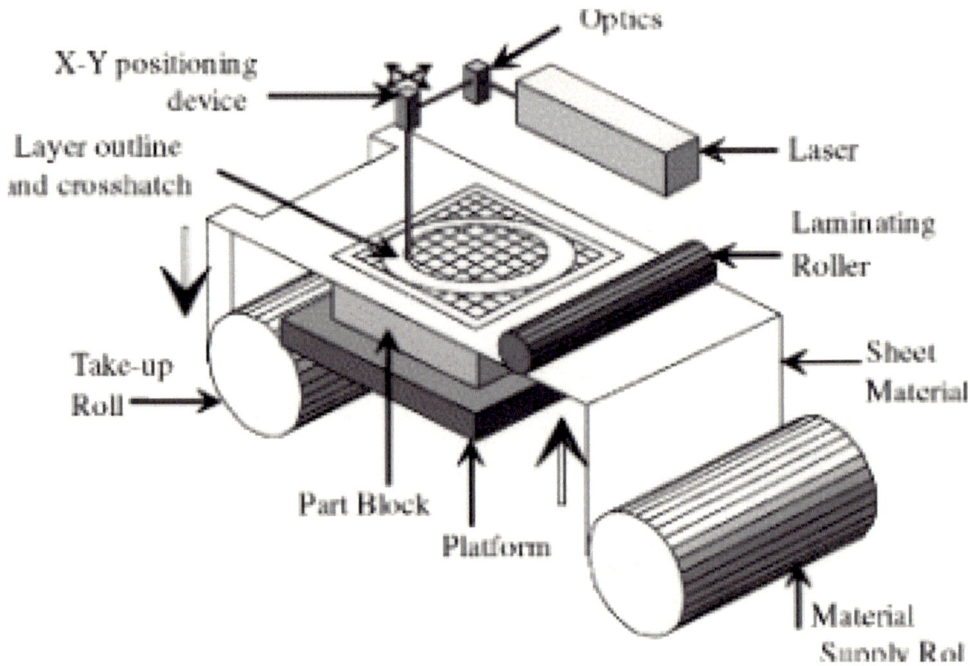

Figure 5.
Laminated tooling process [10].

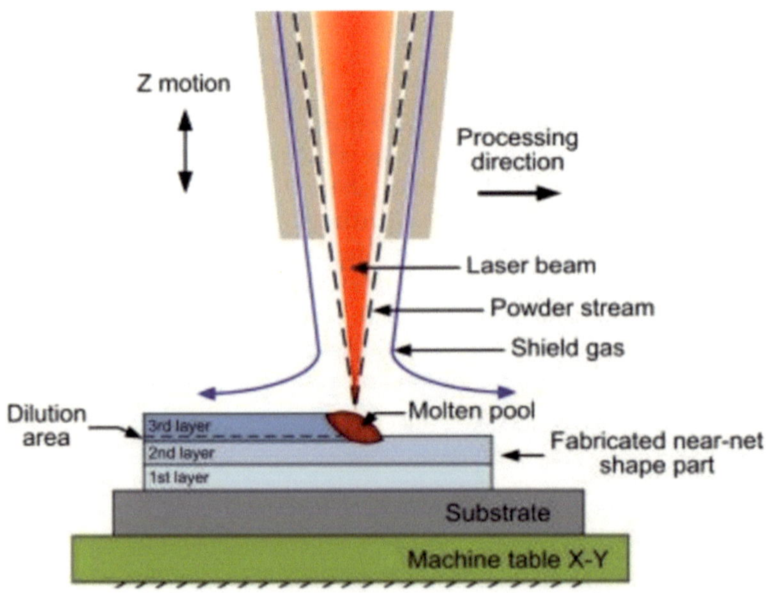

Figure 6.
Laser-engineered net shaping (LENS) process [11].

Technology (IPT) (Aachen, Germany), the fundamental technology was initially created. Three systems were purchased by the business last year.

Components having 100% density are resulted by using this method covering lased cladding and friction. The material is deposited by CMB from a steel wire, and a 1–2 kW HDL laser welds the steel to the workpiece surface. Before every fresh layer is deposited, a cutter with a high speed is used to flat each layer [6].

METHOD	ADVANTAGES	DISADVANTAGES
Direct AIM	Quick process and it produces parts using production thermoplastics.	Low tool strength.
SLS RapidSteel	Quick process, good tool strength and its use for injection molding and die casting	Equipment cost and size limitations.
Copper Polyamide Tooling	Boasts an increase in tool toughness and heat transfer.	Low material strength
EBM Tooling	Improves productivity, reduce cycle time and reduces production cost.	Equipment and powder costs.
DMLS	Suitable production of tool inserts also it has good surface finish.	Low speed requirements.
Laminated Tooling	It has the ability to change the design of parts quickly by the replacement of laminates.	The need for finish machining to remove the stair steps.
LENS	100 percent dense parts	Poor surface finish and small feature definition.
Controlled Metal Build-up	Involves laser cladding and milling that results in 100 percent dense parts	General welding disadvantages.
Prometal	Acceptable results.	Reliability problems.
Direct SL Shell Tooling	Molding cools faster, thus shortening cycle time.	General stereolithography disadvantages.

Table 3.
Advantages and disadvantages of direct methods which use AM.

2.2.9 ProMetal

The ProMetal AM used tool production system of Extrude Hone—named RTS-300—is 3DP method for the manufacture of metal components and tooling. Steel parts up to size of 12 ‘12 ‘10 inches (300 ‘300 ‘250 mm) can be achieved by the machine. ProMetal applications covers vacuum forming, lost foam patterns, injection molding, blow molding, and powder metal part manufacturing [6].

2.2.10 Direct SL shell tooling

Thin SL shells in shell tooling are applied to manufacture inserts reinforced by materials with high thermal conductivity like aluminum-filled epoxy. Therefore, higher mold strengths are attainable in contrast to those achieved by the direct AIM tooling technique, which builds a strong resin mold. Due to aluminum's increased conductivity, which offers faster cooling of mold, the cycle time gets shorter. To improve wear resistance, metal plate can be used to cover the outer surface brief advantages/disadvantages chart of direct methods is depicted in **Table 3** [12].

3. Design of deep drawing tools produced by rapid tooling technologies

In all industries, customized and tailored design is gaining importance, and therefore small series production has increased in the last decade. The increasing number of variant types and also the decreasing number of the same parts affect the manufacturing processes deeply. For instance, metal forming is known to be economical for large series production. One of the main factors affecting the cost of the metal forming process is tool costs. Conventional methods and materials to produce forming tools result in certain disadvantages: Not only the production process takes too much time, but also the whole process is expensive.

Therefore, rapid tooling methodologies are gaining importance also for metal forming technologies in recent years. Rapid tooling methods offer indispensable advantages in time, though the cost of the tool must be optimized according to the number of parts to be produced, material couple to be chosen, and the production methodology.

To optimize the part quality, the method, and the cost prediction, a methodology to predict the most appropriate rapid tooling method as well as the prediction of the tool life becomes indispensable.

As the abrasive wear on the metal forming tools increases, it is getting more and more important to predict the tool wear and the life of the tool, when the tool is in design stage. By this way the most appropriate tool materials, design, and maintenance periods can be planned.

First a methodology to ensure that the chosen rapid tooling technique and material are appropriate to produce the part must be introduced. This includes the dimensional accuracy, mechanical properties, the surface quality of the rapid tooling process, the related production process parameters like deformation of the tool, temperature distribution, and a determination of tribologically matching material couple and surface properties important factors can also be seen in **Figure** 7.

• Surface properties	• Tool loading
• Surface roughness	• Contact pressure distribution on the tool
• Hardness	• Sliding distance
• Young's module	• Temperature distribution
• Plasticity properties	• Friction regimes and friction coefficient
• Anisotropy	• lubrication conditions and regime
• Thermal properties	

Figure 7.
The main factors affecting the choice of the rapid tooling material and production method.

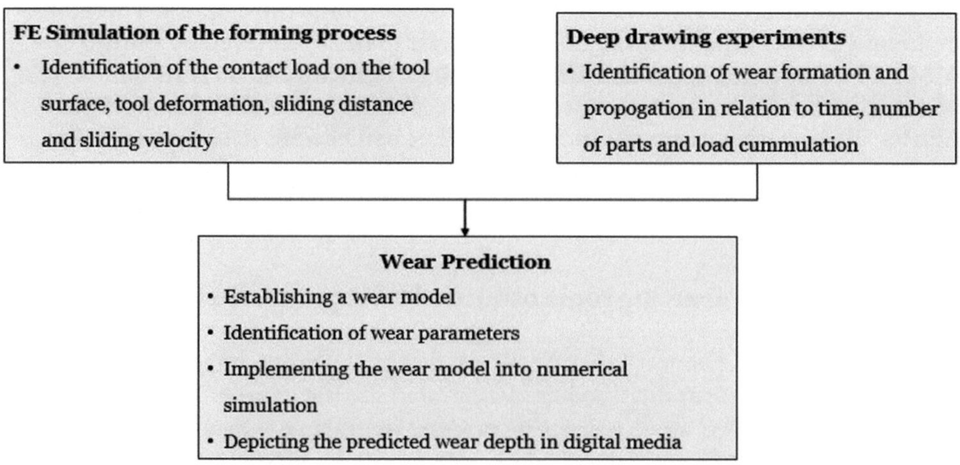

Figure 8.
Approach to predict the wear on forming tools.

Then the number of parts, which can be produced with this rapidly produced tool, must be predicted. In order to predict the operation time of a tool, the following approach is offered.

A metal forming process is economically advantageous, if and only if the tool costs can be controlled and predicted. Forecasts of the number of parts to be produced with the tools are of highest advantage. Therefore, in the PhD thesis, a method to predict tool wear is established and validated and verified an approach is depicted in **Figure 8** [13].

This approach is tested with a simple cup deep-drawing geometry. The die is made of rapid tooling, whereas the punch and blankholder are produced by conventional methods out of tool steel as given in **Figure 9**. For that reason, the punch and blankholder are modeled as rigid body, whereas the die is modeled with 3D deformable elements.

In order to enable testing, a follow-on tool is designed, and an optical and tactile measurement methodology is determined. The die made out of rapid tooling is measured at determined intervals, and the wear propagation over time is measured. Four different locations are measured, and the experiments are conducted up to five times. The details of the repeatability and reliability of the data and measurement methodology can be found in [13].

The contact pressure distribution is obtained by finite element methodology as depicted in **Figure 10**. As the die material is made out of rapid tooling, these materials are generally very susceptible to abrasive wear.

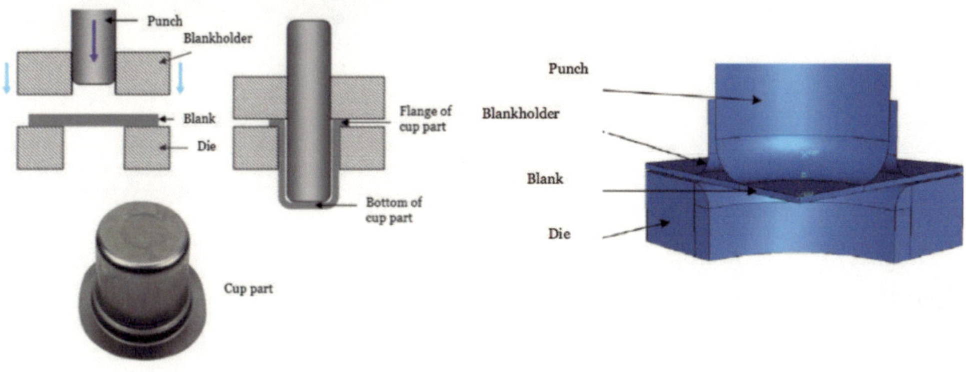

Figure 9.
(a) Schematic presentation of deep drawing of a cup geometry. (b) a quarter model of a simplified circular deep drawing geometry [14].

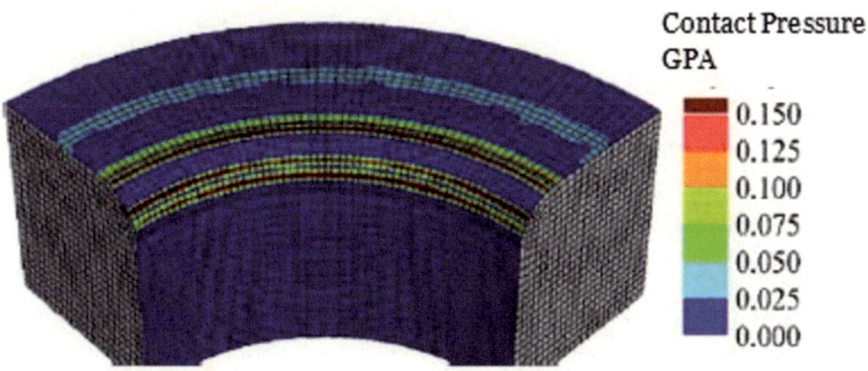

Figure 10.
Comparison of wear behavior of different rapidly produced tools (PA220 (PA-SLS), ZAMAK (SLA + casting), LaserFormA6 (SLS RapidSteel) vs. tool steel (conventional method) [14].

The key technology for predicting the development of wear in metal forming tools is numerical simulation. In forming applications, a wear is commonly described using models based on contact mechanics, the most important one being the Archard wear equation. Here the parameters affecting the wear are contact pressure, sliding distance, hardness, and a tribological constant.

With the offered methodology, the wear depth at each location at each punch stroke can be predicted. **Figure 11** depicts as an example a predicted vs. measured die radius wear rate after 10,000 punch strokes.

The **Figure 12** shows the wear rate of different dies produced by AM technologies, measured at different time intervals. As expected all rapidly produced tools have higher wear, i.e., shorter operation times. LaserFormA6 a kind of stainless

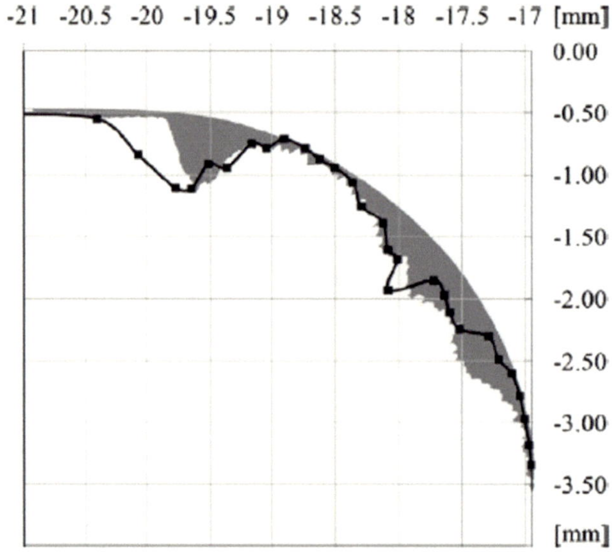

Figure 11.
Contact pressure obtained by finite element simulation [14].

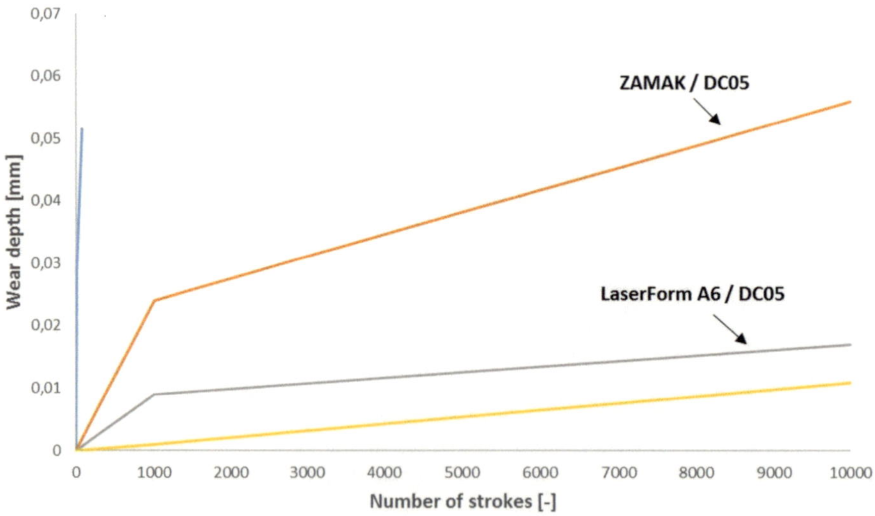

Figure 12.
Measured wear depth in the profile (gray) compared to the simulation results of the same location (black) [14].

steel powder produced by SLS RapidSteel has the best wear resistance as expected, among others. Polyamide die has remarkably high wear rate which indicates that these are not appropriate for any series more than 100 parts. ZAMAK (zinc, aluminum, copper alloy) produced by casting into a SLA mold turns out to have a moderate wear resistance. All these trends obtained in these experiments are in accordance with the Archard theory, where the wear of the tool is indirectly proportional to the hardness of the die materials. Still, when investigated in detail their wear behavior, their elastic deformation, and tribological behavior affect the wear distribution over time, i.e., number of parts, sliding distance, or wear work.

This implies that according to the geometry and the number of the parts needed, this method can also be considered an alternative for sheet metal forming tool.

4. Conclusion

Customized and tailored design is gaining significance in all areas in the recent decades. The growing amount of variants and the declining amount of the same components also have a profound impact on the production procedures. Additive manufacturing is gaining importance due to many advantages, and in tool design it is used mostly due to its lead time advantage. Even if AM in tool production offers indispensable advantages in time, the cost of the tool must be optimized according to the number of parts to be produced, the couple of materials to be selected, and the method of production.

A methodology for predicting the most suitable AM technique for tooling, as well as predicting the life of the tool, becomes indispensable in order to optimize the part quality, process, and price prediction. First, it must be guaranteed that the selected fast tooling method and material are suitable for producing the tool. This involves dimensional precision, mechanical characteristics, surface quality of the selected AM method, associated process parameters such as tool deformation and temperature distribution, and determination of corresponding material pair and surface characteristics tribologically.

In this chapter a metal forming tool in a simplified round die geometry is chosen as an example to predict the wear of a AM-produced tool, and the results show that the AM-produced tool has a shorter operation life, showing a wider range of lifetime depending on the AM technique and the tool material which are used.

From these results it can be concluded that in the chosen AM method, the material must be optimized according to the geometry of the part to be produced, number of parts to be produced, process parameters, tribological requirements, and the time and cost constraints.

Author details

Kıvılcım Ersoy[1*] and Berk Barış Çelik[2]

1 FNSS Defense Systems, Oğulbey, Ankara, Turkey

2 TOBB University of Economics and Technology, Ankara, Turkey

*Address all correspondence to: kivilcim.ersoy@fnss.com.tr

References

[1] Rosochowski A, Matuszak A. Rapid tooling: The state of art. Journal of Materials Processing Technology. 2000;**106**(1-3):191-198. DOI: 10.1016/S0924-0136(00)00613-0

[2] Pham DT, Dimov SS. Rapid Manufacturing: The Technologies and Applications of Rapid Prototyping and Rapid Tooling. Springer Science & Business Media; 2012. 214 p. DOI: 10.1007/978-1-4471-0703-3

[3] Campbell et al. Additive manufacturing in South Africa: Building on the foundations. Rapid Prototyping Journal. 2011;**17**(2):156-162. DOI: 10.1108/13552541111113907

[4] Quail et al. Development of a regenerative pump impeller using rapid manufacturing techniques. Rapid Prototyping Journal. 2010;**16**(5):337-344. DOI: 10.1108/13552541011065731

[5] Evans MA, Campbell RI. A comparative evaluation of industrial design models produced using rapid prototyping and workshop-based fabrication techniques. Rapid Prototyping Journal. 2003;**9**(5):344-351. DOI: 10.1108/13552540310502248

[6] MoldMaking Magazine [Internet]. 2000. Available from: https://www.moldmakingtechnology.com/articles/methods-of-rapid-tooling-worldwide [Accessed: 25 July 2019]

[7] Wimpenny D, Gibbons GJ. Metal spray tooling for composite forming. Journal of Materials Processing Technology. 2003;**138**(3):443-448. DOI: 10.1016/S0924-0136(03)00114-6

[8] Lohfeld S, McHugh PE. Laser sintering for the fabrication of tissue engineering scaffolds. In: Liebschner M, editor. Methods in Molecular Biology. 1st ed. 2012. pp. 303-310. DOI: 10.1007/978-1-61779-764-4_18

[9] Palumbo B et al. Tensile properties characterization of $AlSi_{10}Mg$ parts produced by direct metal laser sintering via nested effects modeling. Material. 2017;**4**:144. DOI: 10.3390/ma10020144

[10] Hague R. The use of stereolithography models as thermally expandable patterns in the investment casting process [thesis]. Nottingham: Nottingham University; 1997

[11] Dehghan A et al. Additive manufacturing methods a brief overview. Journal of Scientific and Engineering Research. 2018;**5**(8):123-131

[12] Rahmati S. The use of stereolithography resins for making tools by the injection moulding process [thesis]. Nottingham: The University of Nottingham; 1999

[13] Ersoy K. Lehrstuhl für umformtechnik und giesseriewesen [thesis]. München: Tecnische Universitat München; 2008

[14] Ersoy K, Nuernberg G, Herrmann G, Hoffmann H. Simulation of wear on sheet metal forming tools—An energy approach. Wear. 2008;**265**:1801-1807. DOI: 10.1016/j.wear.2008.04.039

Chapter 7

Fabrication of Fine-Grained Functional Ceramics by Two-Step Sintering or Spark Plasma Sintering (SPS)

Wallace R. Matizamhuka

Abstract

The majority of functional materials today are based on ceramic materials which find use in a wide range of applications that include magnetic, electronic, optical, thermoelectric (TE) and piezoelectric energy. The properties and reliability of functional ceramic materials are highly depended on the density, grain size and existence of heterogeneities in the microstructure. It is a well-known fact that there is property enhancement at finer grain sizes for most functional materials through a multitude of mechanisms depending on the application. However, what remains a challenge is the success in maintaining fine-grained microstructures using conventional sintering methods. The use of such methods results in uncontrollable grain growth and coarse microstructures which negate the benefits of fine-grained related properties. The use of spark plasma sintering (SPS) technique offers an opportunity to produce fine-grained microstructures with minimum grain growth. However, grain refinement is not always guaranteed during SPS sintering especially under high-temperature sintering conditions. Therefore, sintering conditions that allow densification with minimal grain growth are well suited for microstructural refinement. A modified two-step sintering (TSS) methodology in SPS has proven to yield promising results and has potential use in the production of functional ceramic materials with controlled microstructures.

Keywords: two-step sintering, functional ceramics, spark plasma sintering, functional properties, grain refinement

1. Introduction

Over the past few decades, functional ceramics have played a significant role in advanced technologies owing to their unique thermal, electrical, magnetic, opto-electrical, superconducting and gas-sensing properties. As such, functional ceramics have become the frontiers for advanced technologies such as information technology, medical technology, energy transformation, storage, supply and manufacture technology. For instance, functional ceramics are widely used for electronic applications as they can operate at high power and high frequencies, at high temperatures and harsh conditions. Their capability to combine properties such as electrical insulation and magnetism, which is not possible with metals, gives them an additive advantage.

Functional ceramics are produced from chemically synthesised powders in the form of oxides, nitrides, carbides and borides mainly through a powder metallurgy route. The properties of functional ceramics are microstructure sensitive, and microstructural features such as grain size, composition, homogeneity and grain boundary constituents are critical to their performance and reliability. The processing route dictates the final microstructural features obtained; thus, the choice/design of a processing route is key in material functionality. For instance, fine grain size has been experimentally proven to amplify functional material properties such as electrical conductivity, thermal conductivity, piezoelectric and ferroelectric properties [1]. It must be mentioned though that the fabrication of dense nanostructured functional ceramics by conventional sintering methods is quite challenging owing to the uncontrollable high grain growth rates [2]. This explains the shift in research focus towards nanostructured functional materials in the past few years.

In recent years, spark plasma sintering (SPS) technology has proven its capability to fabricate fine-grained microstructures possessing superior properties for a wide range of materials [2]. This method is increasingly being applied in the production of functional ceramic materials. It is against this background that the present chapter is aimed at giving an insight on the progress made so far and, furthermore, how the resulting microstructures and properties align with the required functions. It is imperative that a background on the various applications of functional ceramics be given prior to a detailed discussion on the SPS sintering methodologies.

2. The most popular functional ceramics and their applications

Functional ceramics are materials tailored to possess exceptional properties (electrical, thermal, optical, piezoelectric and magnetic properties) by controlling the composition and microstructures [3]. These materials are being utilised in a broad range of applications owing to the distinct advantages they offer in comparison to metals. The list below is not meant to be exhaustive but to give a qualitative review on the applications of the most popular functional ceramic materials.

2.1 Piezoceramics

Piezoelectric ceramic materials couple electrical and mechanical responses in their functioning and are widely used for electromechanical sensors and actuators. These materials normally produce an electrical response in the form of either a voltage or charge proportional to the applied stress when subjected to a mechanical force. Conversely, an applied voltage can be converted into mechanical energy such as in piezoelectric motors and sound-/ultrasound-generating devices. Piezoelectric materials are widely used in dynamic applications which include mechanical impact, ignition systems, vibration suppression and sensing [4]. Typical examples of piezoelectric materials include crystalline quartz, barium titanate ($BaTiO_3$), vanadium niobate and lead zirconate titanate (PZT) [3]. In recent years, research focus on lead-free piezoelectric materials has been intensified aimed at replacing lead-based materials in electronic devices for the sake of human health and preservation of the natural environment [5].

Piezoelectric materials are produced as multilayered components consisting of electrode-ceramic stacks which can be simple/complex shapes. Various techniques have been developed to fabricate the piezoelectric ceramics without conducting post-processing. These include injection moulding [6, 7], embossing [8] and fused deposition method [9]. The powder injection moulding (PIM) process has received much attention owing to its ability to produce complex-shaped, micro-sized PZT components with minimum damage to the sintered ceramic [10]. The multilayered components are subsequently co-sintered at temperatures less than the melting point of the electrodes typically 1200–1300°C. However, lower sintering temperatures are preferred to avoid damaging the inner electrodes in the stack. It has been observed that the piezoelectric coefficient which directly influences the performance of piezoceramics is strongly influenced by the grain size [11]. Despite the extensive studies carried out over the past decades on the grain size effects on the physical properties of these materials, there are still major controversies on the dependence of piezoelectric and ferroelectric properties on the grain size [12]. There are a number of discrepancies in the existing literature which will be discussed later in this chapter.

2.2 Magnetic ceramics

Magnetic ceramic materials are extensively used in electronics and information communication fields [13]. They are generally classified as 'soft' and 'hard' magnets where soft implies large magnetic fields cannot be generated on the outside, whereas in the case of hard magnets, a magnetic field is generated around the magnet itself. Two broad groups of materials are widely used in the industry, i.e. metal magnetic materials and complex oxide containing trivalent iron ion (ferrites) magnetic materials (referred to as ceramic magnets).

Magnetic materials are generally used in the form of multilayer core of rolled thin plates or in the form of dust core [13]. At high frequencies, most metallic magnets tend to lose their magnetic properties (permeability and magnetic flux density) due to low electrical resistivity. On the other hand, ferrites (ceramic magnets) show higher electrical resistivity and smaller eddy current loss at high frequencies; hence, they are more widely used in alternating magnetic fields in comparison to metal magnetic materials. The hard ferrite is used extensively as permanent magnets for speakers and motors. One of the critical magnetic characteristics required for high-frequency materials is high permeability and is defined as the ratio between the magnetic flux density, B, and magnetic field, H, as follows:

$$\mu = \frac{B}{H} \tag{1}$$

Permeability is a structure sensitive characteristic and is strongly affected by the microstructure of the sintered material. There are two general compositions used for oxide magnetic materials, spinel type ($MeFe_2O_3$) and garnet type ($Me_3Fe_5O_{12}$); typical examples include $MnFe_2O_4$ and $Y_3Fe_5O_{12}$, respectively. Owing to the complex compositional nature of these oxides, a powder metallurgy route is normally employed for the production of oxide magnets. The microstructure and compositional control are quite critical elements of the magnetic properties of the final products. Further, magnetic properties of materials have been shown to change from those of multidomain to those of single-domain structure as the grain size is reduced below a critical size [14]. The introduction of fine-grained sintered magnetic materials has opened some opportunity for new potential applications as well as complexity on basic research [14].

2.3 Dielectric ceramics

In the last few decades, the rapid development of modern communication devices such as cellular telephones, antennas and global positioning systems has energised research in microwave dielectric materials [15]. Dielectric ceramics are materials used widely in advanced electronic devices such as capacitors and microwave resonators. They are classified into two broad groups based on their dielectric properties. High-quality factor materials are characterised by linear changes in polarisation with applied electric field. This group is dominated by titanate-based materials which normally sinter at temperatures higher than 1100°C; typical examples include TiO_2, $MgTiO_3$, $CaTiO_3$ and $SrTiO_3$ [13]. This group is characterised by a dielectric constant ε_r of less than 1000. The second group is characterised by materials possessing a dielectric constant ε_r higher than 1000. Typical examples include $BaTiO_3$-based dielectric and lead-based dielectrics.

Ceramic capacitors are widely produced as sintered thin plates in a reducing atmosphere (low P_{O_2}). In previous studies, the particle size effects of $BaTiO_3$ on dielectric properties have been carried out with several models of the critical size of ferroelectricity being proposed. Reliability study results have shown that the dielectric layer should be pore-free with fine grain sizes (typically 0.8 μm) for enhanced performance [13].

2.4 Thermoelectric ceramics

Thermoelectric (TE) ceramic materials can directly convert heat energy to electric energy due to thermoelectric effects [16]. TEs provide an alternative environmentally friendly energy conversion technology which is compact, high reliability, has no pollutants and is feasible over a wide temperature range. The majority of thermoelectric devices operating near room temperature are based on Bismuth telluride (Bi_2Te_3) and its alloys. These materials have been produced by a variety of methods which include powder metallurgy techniques such as hot pressing (HP), SPS, Bridgman and zone melting and high-pressure sintering methods. Recent studies have shown that grain refinement of Bi_2Te_3-based alloys can greatly enhance thermoelectric performance [16]. The performance of thermoelectric materials is based on a dimensionless figure of merit (ZT) as follows:

$$ZT = \frac{S^2 \delta T}{\kappa} = \frac{S^2 T}{(\kappa_e + \kappa_l)\rho} \tag{2}$$

where S, δ, κ, ρ and T represent the Seebeck coefficient (S), electrical conductivity (δ), thermal conductivity, resistivity and absolute temperature, respectively.

Thermal conductivity of TE materials consist of two parts: lattice thermal conductivity (κ_l) and electronic thermal conductivity (κ_e). In principle, a high ZT is obtained by large values of both seeback coefficient and electrical conductivity, while thermal conductivity (κ) is minimised to maintain the temperature difference (T) producing the Seebeck coefficient [17, 18]. However, this requirement contradicts the Wiedemann-Franz law which requires the electronic part of thermal conductivity to be proportional to electrical conductivity, and the Pisarenko relation limits the simultaneous enlargement of α and δ [19]. This makes it difficult to enhance the ZT using the tuning of carrier concentration alone.

Over the years, a number of strategies have been adopted to enhance the power factor and reduce thermal conductivity of TEs. This has resulted in the development of three generations of TEs over the 200-year period since their discovery in 1821. The development history has been characterised by achieving high ZTs > 2.0 through new concepts and technologies. The first TE generation devices are characterised by ZT ~ 1.0 operating at power conversion efficiencies of 4–5% [17]. In the 1990s the introduction of nanostructures increased the ZT values by about 70% to ZT ~ 1.7, and the power conversion efficiencies can be expected to be 11–15%. In the third generation, some new concepts such as band structure engineering by doping, reduction in lattice thermal conductivity, nanostructuring and all-scale hierarchical architecturing and quantum confinement effects have been introduced to enhance Seebeck coefficients [16].

3. Processing of functional ceramic materials

It is apparent that the increasing demand for ceramic materials in more advanced technological applications has resulted in greater need for improved properties and reliability of functional materials [2, 13]. The fabrication process plays a critical role in final material characteristics. In other words, the properties of ceramic materials are dictated by the microstructure which is a function of the processing method utilised. Thus microstructures can be tailored through fabrication processes to produce desired properties. In the past few decades, there has been a wide acceptance among powder metallurgists that the quality and reliability of ceramic materials are largely dictated by utilising powders of controlled purity, particle size and size distribution, shape and degree of agglomeration. The characteristics of starting powders are determined by their production method of which a variety of methods are available for the production of ceramic materials. The processing methods are broadly classified into solid-state processing (e.g. mechanical alloying, self-propagating high-temperature synthesis (SHS), laser ablation) and solution chemistry (e.g. sol-gel, polymer pyrolysis, hydrothermal methods) [2]. The chemical processing methods are generally more expensive than solid-state methods but offer more strict control of the powder characteristic [13]. The choice of a powder processing route will therefore largely depend on the production cost and its capability to achieve desired powder characteristics.

Chemical methods involving chemical reactions under carefully controlled conditions normally result in ultrafine nanometric powders (<100 nm) with a narrow particle size distribution [13]. The main attraction in using nanometric powders is their ability to sinter at lower temperatures (typically <0.5 Tm); this is in accordance with Hering's law discussed in the next section. The diffusion distance during sintering is drastically shortened in nanostructured powders. Moreover, an enhancement of material properties is expected owing to a reduction in the flaw size, and a higher density of highly disordered interfaces is also attained at nanometric particle size range. On the other hand, powders produced by mechanical methods possess a wide particle size distribution which may lead to higher packing density in the green body. However, this advantage is far outweighed by the difficulty in microstructural control during sintering as large grains grow uncontrollably at the expense of the smaller grains, thus making grain size control impossible. However, it is important to underline that as particle size decreases, below ~0.5 μm, particles become more difficult to handle and tend to agglomerate resulting in nonuniform consolidation of powders. Thus the use of nanopowders requires proper control and handling to ensure high-quality properties are attained in the final products.

4. Spark plasma sintering technology

In 1906, Bloxam filed the first patent on the successful consolidation of powder using the SPS technology [20]. Steady progress was made in the mid-1980s into the 1990s. The SPS technology sinters in a conducting die with a simultaneously applied mechanical pressure and DC pulses which allows for simultaneous densification and sintering process [21]. The SPS set-up consists of a graphite die filled with powder feedstock, uniaxial hydraulic pressing device which achieves 50–250 kN and an electric pulse current of low voltage (<10V) and high currents (1–10 kA) [2]. The system has achieved heating rates of up to 1000°C/min which makes it possible to sinter over very short durations [22]. It can be operated under vacuum or inert gas atmosphere at atmospheric pressure with a maximum temperature of 2400°C.

The sintering mechanisms in SPS are a result of three effects, namely, mechanical, thermal and electrical [2]. The fast heating rates achieved in SPS enables densification while retarding microstructure coarsening owing to the short times required to reach sintering temperature. This allows for the densification of nanopowders with minimal grain coarsening [22]. The SPS system offers a number of advantages over the conventional sintering systems such as hot pressing, hot isostatic pressing (HIP) which include high sintering speeds, high reproducibility, better control of sintering energy and reliability.

The mechanism of sintering is not well understood but several authors have postulated a number of theories. The widely accepted SPS sintering mechanisms involve joule heating, plasma generation and electroplastic effect [2]. The electrical effects are a function of the electrical properties of the powders. For powders that are electrically conducting, current can easily flow through, and heat is generated mainly by joule heating and transferred to the bulk of the powder by conduction (see **Figure 1**) [23]. In the presence of an applied pressure, the electric current through the particles enhances formation of interparticle bonds through localised welding, vaporisation or cleaning of powder surfaces [22]. This ensures a smoother and more favourable path for the current flow. This also promotes the production of high-quality sintered compacts at lower temperatures in a shorter time than conventional sintering methods. The sintering of nonconducting powders, although not well understood, is thought to occur through grain boundary migration and matter transport at higher input voltages.

Although the SPS has the capability to sinter at high heating and cooling rates, the expectation is that the system can sinter without appreciable grain growth. However in reality this is not always the case; a complete avoidance of the grain growth at the sintering temperatures for most nano-grained materials will always promote grain growth. It is therefore imperative to adopt an approach/methodology

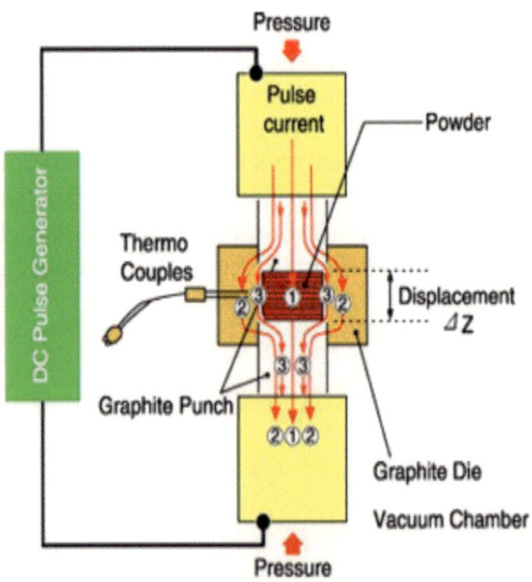

Figure 1.
Pulse current flow through the spark plasma sintering technology [22].

that is more effective in minimising grain growth using the SPS system. The *two-step* sintering (TSS) approach has been found to be effective in minimising grain growth of ceramic materials during sintering. It is thus important to dedicate the following section to the success studies on the sintering of functional ceramics to give an insight and a better understanding of the TSS method.

5. Two-step sintering methodology

The reliability of ceramic materials is a key function, and it dictates their ultimate performance. A carefully controlled microstructure has a greater impact on the properties and reliability of functional ceramic materials. In the previous section, it has been shown that the development of highly dense nanometric or ultrafine-grain-sized ceramics is not easily achievable through conventional sintering. Although the SPS technology has shown great potential in the production of highly dense nanometric materials, it is difficult to maintain microstructural refinement under the high sintering temperatures. It must be underlined that solid-phase sintering requires high temperatures to facilitate diffusion which promotes material densification. However diffusion processes promote not only densification but also grain growth [24]. To achieve grain refinement during sintering, it is therefore imperative to develop a sintering methodology which promotes only densification without stimulating grain growth. This method has been improved over the years to achieve better microstructural refinement. The so-called two-step sintering (TSS) was subsequently introduced in the 1990s by Chu et al. [25]. In essence, the technique consists of two stages of consolidation process, i.e. a first stage performed at relatively low temperature followed by a higher-temperature stage and subsequent cooling.

However, the higher-temperature stage if not adequately controlled can lead to some grain growth. In 2000, Chen and Wang proposed a modified TSS methodology which effectively suppresses the accelerated grain growth in the second stage [24, 26]. In the modified TSS approach, a high-temperature heating is performed first for a short duration followed by structural freezing and sintering at a lower temperature. The idea of heating to a higher temperature (T1) followed by fast cooling with no sintering holding time (stage 1 in **Figure 2**) is to eliminate residual porosity at higher temperature and develop a network of grain boundary anchoring at triple points [11].

These anchored triple points are thought to have higher activation energy for matter migration than the grain boundaries. The second step effectively proceeds in

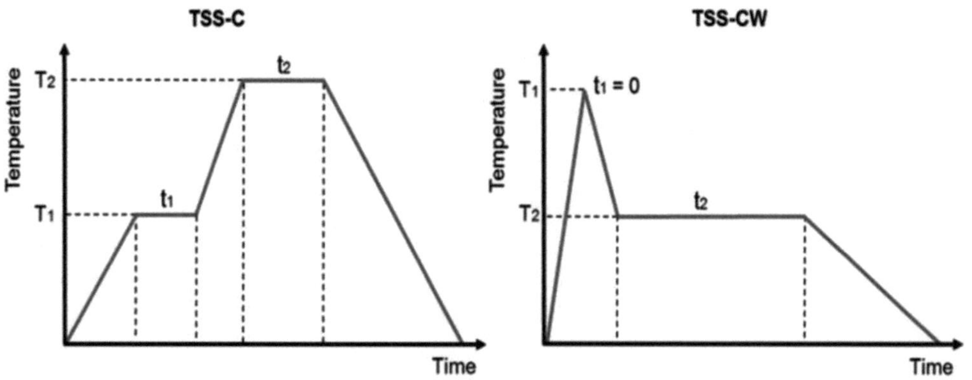

Figure 2.
Schematic illustration of the differences between the two TSS approaches [11].

a *frozen* microstructure due to slower kinetics [27, 28]. The *kinetic window* separates grain boundary diffusion and grain boundary migration. Grain growth experienced at intermediate and high temperatures in fine-grained materials is driven by the significantly higher capillary pressures available in ultrafine grains. For instance, for a grain boundary energy (and surface energy) in ceramics of 1 J/m^2, a capillary pressure of the order 20 MPa at 100 nm, 200 MPa at 10 nm and 2000 MPa at 1 nm grain size. This implies there are significantly higher pressures locked in ultrafine grains to ignite additional kinetic effects at elevated temperatures. Thermodynamically, the temperature (T2) is sufficiently high to allow grain boundary diffusion with minimal grain boundary migration; this promotes densification without significant grain growth. It is however important to select the most suitable T2 temperature. In the case that T2 is too low, sintering proceeds for a prolonged period until it becomes exhausted; on the other hand, if T2 is too high, grain growth is likely to occur.

5.1 Two-step sintering as applied to functional ceramic materials

The use of the TSS methodology to obtain ceramic materials of controlled microstructure has become standard practice. This section gives a detailed discussion on the effects of processing characteristics on grain density and size as well as their contribution to the improvement of mechanical properties of a number of functional ceramic materials investigated in previous studies.

The pioneering work of Cheng and Wang in 2000 serves as the beginning of a new era in TSS methodology. In one of their successful studies, Chen and Wang obtained a density of 99% and a grain size of 123 nm using a T1 temperature of 1250°C and T2 temperature of 1100°C for a pure Y_2O_3 ceramic material [24, 26]. After several experimental studies, Chen and Wang concluded that the success of grain growth suppression in their work was mainly attributed to triple-point immobility irrespective of whether doping agents were used or not [11]. In a separate study, Mazaheri et al. [29] obtained dense samples of ZnO with limited grain growth under varying conditions. The starting particle size of the ZnO was 31 nm. The most interesting result was obtained with a T1 of 800°C and T2 of 750°C; a relative density and grain size of 98% and ~68 nm were obtained, respectively [29]. The same authors proved that slightly higher temperatures (850°C and 780°C for T1 and T2, respectively) resulted in grain growth and a lower densification of 86% using ZnO material [29]. A further study done at even higher temperatures, with a starting ZnO powder of grain size of 400 nm (0.4 μm) and a T1 of 1100°C and T2 of 1050°C, resulted in a relative density of 95.1% and a grain size of 3.9 μm, signifying the ineffectiveness of the TSS methodology at higher sintering temperatures and larger particle sizes. Several other TSS studies carried out on the ZnO material proved that the use of dopant agents such as Bi_2O_3, Sb_2O_3, CoO and MnO assisted in suppressing grain growth [11].

Yttria-stabilised ZrO_2 (YSZ) is one of the most important functional ceramics which find its use in a wide range of applications. Several attempts have been focused on attaining nanometric YSZ materials to improve its functional properties. Mazaheri et al. [30] obtained fully densified 3YSZ with an initial grain size of 75 nm and a pressureless sintering regime of T2 (1150°C) and T1 (1300°C) with an isothermal holding time of 30 h at T2 and 1 min holding time at T1. The final grain size achieved was 110 nm. In a separate study, Suarez compared SPS sintering with pressureless TSS methodologies. The starting material was a three Y_2O_3-stabilised tetragonal ZrO_2 (3YTZ) with an initial average particle size of 65 nm. The TSS methodology used a pressureless sintering regime as T1 at 1350°C without holding time and T2 at 1200°C for 15 h, and a final grain size of 125 nm was obtained. On the other hand, the SPS method was carried out at 1150°C with a heating rate of 300°C/min and an isothermal holding time of 30 mins at a pressure of 150 MPa.

The grain size obtained was 115 nm. Several other modifications as summarised in **Table 1** were carried out, and an observation that was common to the majority of the studies is that the homogeneity of the green body was very critical to the success of the sintering method in its effectiveness of grain growth control [11].

As mentioned earlier, piezoelectric materials' performance is strongly influenced by the grain size of the constituent particles making up the ceramics. High-performing piezoceramics have been obtained using the TSS approach. A number of studies have shown that lower sintering temperatures using the TSS methodology can be utilised to obtain ultrafine grain sizes. $BaTiO_3$ is one of the most popular piezoceramics studied, and a number of studies have been carried out to obtain fine-grained microstructures. In their work, Kim and Han [31] used a 1% dysprosium (Dy)-doped $BaTiO_3$ with a particle size of approximately 17 nm which was compacted at 300 MPa at room temperature. The TSS profile used a T1 of 1300°C and a T2 of 1100°C with a 20 h holding time. A grain size close to 1 μm was obtained at a relative density of 95%. In a separate study, Wang et al. [32] studied two different piezoceramic composites, i.e. pure $BaTiO_3$ with a particle size between 10 nm and 30 nm and a nanometric ferrite of composition $Ni_{0.2}Cu_{0.2}Zn_{0.6}Fe_2O_4$ with a starting particle size of 10 nm. The two powders were compacted isostatically at 200 MPa. The two piezoceramics were sintered using two different sintering programmes, i.e. a T1 of 950–1250°C for $BaTiO_3$ and at 850–930°C for ferrite; a cooling rate of 10°C/min was used in both cases. The samples were cooled at 30°C/min to a

Sample	Method	Starting grain size (nm)	Pressure (MPa)	T1 profile (°C)	T2 profile (°C)	Relative density (%)	Grain size	Ref
Pure Y_2O_3	TSS	10–60	—	10°C/min to 1250	50°C/min to 1100, 6–30 h dwell	99	123 nm	[24, 26]
Pure ZnO	TSS	31	—	800	750	98	680 nm	[29]
Pure ZnO	TSS	400	—	1100	1050	95.1	3.9 μm	[34]
3YSZ*	TSS	0.27 μm	—	1500 (5 min) 10°C/min	1300, 10 h	—	0.59 μm	[35]
3YSZ	TSS	75	150	1300, 1 min	1150, 30 h	Density at T1 (83)	110 nm	[30]
3YSZ	SPS	65	150	—	1150, 150 MPa, 300°C/min, 30 min	—	115 nm	[36]
3YSZ	TSS	65	—	1300	1200, 15 h	—	—	[36]
3YSZ	TSS	60–120	—	1300, 10°C/m	1175, 20 h	99.2	184 nm	[37]
3YSZ	H-SPS	60–120	100 at 600°C, 3 min	300 MPa at 1000, 5 min	1175, 300 MPa, 30 h	97.4	173 nm	[37]
8YSZ	SPS	58	50, 15 min	1150, 200 °C/min, 20 s, 10 MPa	1050, 50 MPa, 2 h	99.8	190 nm	[38]

3YSZ, 3 mol% Y_2O_3-stabilised ZrO_2.

Table 1.
A summary of the two-step methodology used to produce different materials.

T2 range of 1150–850°C for $BaTiO_3$ and 750–870°C for the ferrite with an isothermal holding time of 20 h. For the $BaTiO_3$, the best result achieved showed a grain size of 35 nm and was sintered at 950°C in T1 and at 900°C in T2 for 2 h. On the other hand, a grain size of 200 nm was achieved for the ferrite sintered at 850°C in T1 and 800°C for 6 h in T2. In comparison to the result of Kim and Han discussed earlier, the use of a lower sintering temperature enabled the formation of $BaTiO_3$ with a finer grain size. To elucidate this point, Karaki et al. [33] obtained a mean grain size of 1.6 μm with a relative density of 98.3% starting with a $BaTiO_3$ of average particle size of 100 nm, cold compacted at 200 MPa. A TSS profile almost similar to that of Kim and Han et al. consists of a heating rate of 10°C/min to T1 (1230–1340°C) held for 1 min and a cooling rate of 30°C/min to T2 (1150–1200°C) held for 2, 4, 5 and 20 h. The best result was obtained with a sintering profile consisting of T1 (1320°C) and T2 (1150°C) with a 15 h holding time. There is a clear indication from the above results that $BaTiO_3$ ceramics with high sintered density are obtainable at lower sintering temperature (≤1000°C) for both T1 and T2. Some of the authors observed that the piezoelectric coefficient is strongly influenced by the grain size [11]. Nanometric $BaTiO_3$ ceramics possess superior piezoelectric coefficients in comparison to their micrometric counterparts.

As mentioned earlier, the current work is not meant to be exhaustive but to give a qualitative insight on the research covered so far. **Table 1** gives a summary of some of the successful work carried out and highlights some critical aspects in the TSS methodology as a grain refinement process.

Specific particularities were observed in each of the sintering cycles above. The majority of the authors compared their results with those obtained by conventional sintering and the TSS methodology resulted in superior grain refinement. The initial powder features such as particle size, microstructural homogeneity and green density are quite critical in the success of the TSS process. In the majority of cases, the difference between the T1 and T2 temperatures is <1500°C, and a larger holding time in T2 allowed smaller grain sizes to be obtained. The TSS method has also been shown to improve the material properties (both mechanical and physical).

In some cases the use of dopants was effective in inhibiting grain growth. Although the TSS approach has shown great success in effecting grain refinement, the very long isothermal holding times at T2 might not be suitable for commercial purposes. The SPS technology offers an alternative route for grain refinement, and results in **Table 1** show that it is more effective for materials such as YSZ ceramics. The SPS process generally can achieve grain refinement over shorter time periods.

6. The use of SPS technology in the synthesis of functional ceramic materials

SPS technology provides an alternative and more effective route for grain refinement of ceramic materials. As discussed earlier, conventional sintering is ineffective in refinement of ceramic materials owing to the excessive grain growth of fine powders at elevated temperatures [20–22]. The TSS methodology was developed to mitigate this problem, and it provides a low-cost and effective route for grain refinement. However, this method requires very long isothermal holding times to effect sintering without grain growth. Therefore, it might not be suitable for high production rates (commercial purposes); in addition the prolonged holding times are likely to increase the energy costs. Moreover, there are ceramics that require pressure-assisted sintering to impart the required strength to the component during sintering. The shortcomings above can be minimised with the use of SPS technology which has been briefly discussed in the previous section.

The SPS technology has been successfully used in the production of nanometric functional ceramic materials. The discussion below presents some examples where SPS has been utilised to obtain bulk ceramic nanomaterials. The discussion is not meant to be exhaustive but to impart some critical in-depth knowledge on the synthesis procedures of a few selected case studies.

6.1 SPS of nanostructured magnetic ceramics

The magnetic properties of ferrites (magnetic ceramics) are structure sensitive and are affected by a number of factors such as phase composition, crystallite size and shape and the quantity of heterogeneities [39]. It is therefore critical to use a synthesis method capable of producing superior magnetic properties through control of structural homogeneities. The use of SPS technology is thought to promote ordering of spatial positions of magnetic moments of metal ions in the composite crystal lattice through pressure-assisted sintering [40]. The result is the formation of new magnetic phases through a pressure-induced transformation of the nanocrystalline phases [39]. It must be noted however that the mechanism of this transformation in SPS is not well developed for most of the ceramic ferrites.

Papynov and co-workers studied the magnetic properties of nanostructured ferrites using SPS technology (α-Fe_2O_3 and α-Fe_2O_3-Fe_3O_4 composite) [39]. The authors established that the value of magnetisation increases significantly with increasing sintering temperature and reached a value of 10.2 emu/g at 1100°C (equivalent to a tenfold increase). This was attributed to changes in the crystalline phase and to a lesser extent growth of ferrite grains which may affect magnetisation.

In their work Gaudisson et al. [41] consolidated a nanosized magnetic powder into a high-density solid at 750°C for 15 min to a final grain size range of 150 nm. In a separate study, a nanostructured Co-ferrite was shown to be sensitive to heating rate under the same sintering temperatures and times in SPS. A higher heating rate (80°C/min) maintained a finer grain size of 70 nm than a lower heating rate (15°C/min) which produced a grain size of 290 nm for two powders which were processed at the same sintering temperatures and times (2 min at 600°C followed by 5 min at 500°C) [41]. Ultrafine, highly dense yttrium iron garnet (YIG) was produced by SPS treatment at 750°C for 15 min at 100 MPa in wide contrast to the typical parameters used in conventional sintering which requires higher sintering times (typically 1350°C for a few hours) [42].

A $SrFe_{12}O_{19}$ hexaferrite with a grain size of 400 nm was obtained using SPS at 1100°C for 5 min with a maximum density of 5.15 g/cm^3. There was a big contrast with conventional sintering at 1240°C for 2 h which produced a density of 4.83 g/cm^3 and a grain size double that of the SPS-produced material [43]. Harder magnetic properties were obtained from the SPS-produced ferrite. In another study, harder magnetic properties were obtained for an SPS-sintered Ba-hexaferrite owing to limited grain growth; grains of 100–150 nm were obtained in comparison to conventional sintering which produced a grain size of 1.5–8 μm [44].

6.2 SPS of nanostructured piezoceramics

The grain size effect on the macroscopic functional properties of piezoceramics has been widely researched. There are however very few studies that correlate grain size and property stability. It has been shown that improved performance, high permittivity miniaturised devices can be obtained by microstructural control such as grain size and homogeneity [45]. Arlt and co-workers have shown the strong dependency of $BaTiO_3$ functional properties on the microstructure and grain size [46]. Moreover, large grain sizes are detrimental to the mechanical strength of ceramic-based devices.

Several studies have been dedicated towards investigating the effect of grain size on the piezoelectric properties of $BaTiO_3$ ceramics down to nanometric scale.

$BaTiO_3$-based piezoceramics is one of the most studied using SPS. It has been demonstrated that SPS technology is effective in stabilising the metastable $BaTiO_3$ cubic phase and reducing the intergranular effects on permittivity and DC resistance [45]. Moreover, SPS samples have shown higher permittivity values typically below the Curie temperature (Tc) [47]. It has been demonstrated that at finer grain sizes, the dielectric constant at the transition temperature decreases and Tc shifts to lower temperatures [45].

Lead-based piezoceramics have dominated the market of piezoelectric ceramics for a long time. However, their continued use is now questionable owing to the associated health risk especially during processing. Another major concern in the sintering of PZT piezoceramics ($Pb(Zr,Ti)O_3$) is the high sintering temperatures which promote the vitalization of lead [48, 49]. Moreover, a number of the proposed alternative piezoceramic materials also contain highly volatile elements such as in $(Na,K)NbO_3$ which makes their sintering ability quite poor. The use of SPS has enabled suppression of lead loss through rapid heating rate, lower sintering temperature and shorter sintering times [50]. In one study, Han et al. demonstrated that the use of SPS can lower the sintering temperature of a $Pb\ (Zr_{0.52}Ti_{0.42}Sn_{0.02}Nb_{0.04})O_3$ piezoceramic by a substantial 200–300°C while maintaining a high relative density (>99%) [51]. In a separate study, a $(Na_{0.535}K_{0.485})_{1-x}Li_x(Nb_{0.8}Ta_{0.2})O_3$ ($x = 0.02$–0.07) ceramic with improved mechanical and electrical properties was produced using SPS method [52].

There is an assumption that the nonlinear response of piezoceramics is grain size dependent; this is understood to be the variation of functional properties under an external stimulus. The two major contributors to nonlinear response of piezoceramics are the intrinsic (i.e. the contribution of composition, crystal structure, etc.) and extrinsic (i.e. grain size, domain wall dynamics, etc.) contributors [53, 54]. This implies that a significant decrease in grain size has the potential to produce a notable modification of nonlinear response in piezoceramics. It therefore means the stability of piezoelectric properties may be improved by controlling the grain size.

6.3 SPS of nanostructured thermoelectric ceramics

The wide application of TEs has not been realised mainly owing to low conversion efficiencies. For instance, commercially available TE materials possess a low ZT of 1 and average conversion efficiency of ~5% [55]. In order to promote the practical applications of TEs, it is critical to synthesise TE materials with ZT values >1; a TE device with ZT = 3 operating between room temperature and 773 K would yield ~ 50% of the Carnot efficiency [56]. It is evident from previous reviews that the key strategy in the improvement of ZT values for TEs has been the increase in the seeback coefficient and reduction in thermal conductivity. However, no significant improvement in ZT values has been reported through the tuning of these properties. Theoretical predictions have shown that nanostructuring can enhance the seeback coefficient through modification of density of states and can reduce the thermal conductivity by selective scattering of phonons, resulting in good ZT values. It should be noted here that the TE properties of nanostructured materials also depend on the size and morphology of microstructural features; thus, microstructural engineering is key in the development of TE materials. In 2005, Yu et al. observed that the seeback coefficient and thermal and electrical conductivities are all significantly dependent on grain size; this was confirmed on $CoSb_3$ TE materials [57].

It has been proven that the main design principle for the future TEs is the use of nanostructured architectures. A number of approaches have been utilised in developing nano-inclusions which are effective in reducing the lattice thermal conductivities [16]. Such methods include in situ dispersion of partially oxidised nanoparticles in matrix [58], endotaxial nanoprecipitates [59, 60] and embedded nano-inclusions [61, 62]. The SPS technology has been utilised in fabricating highly dense and fine-grained TEs [63]. Nanocomposite grains are believed to be effective in scattering phonons with a broad wavelength which enhances the functional properties of TEs [55]. By nanostructuring a wide variation in ZT values ranging from 0.4 to 1.7 has been obtained for nanocomposites with similar composition. A ZT of about 1.5 at 390 K was achieved for a $(Bi,Sb)_2Te_3$ nanocomposite produced by a combination of melt spinning of single elements followed by SPS sintering. Another $Bi_{0.52}Sb_{1.48}Te_3$ nanocomposite material had a ZT ~ 1.56 [64, 65]. It has also been proven that by combining mechanical alloying and SPS sintering, one can achieve high ZT of 1.5 at 700 K in $AgPbmSbTe_{m+2}$ nanocomposite [66].

Most of the bulk TE materials with highest ZT values are fabricated through the SPS process. Bi_2Te_3 compounds have been produced with ZT values ranging from 0.7 to 1.8 in the SPS [67]. The reason for a wide range of ZT values has been attributed to varying initial green densities which is key in determining the inner temperature of the sample. Moreover, powder aggregates can lead to inhomogeneous distribution of temperature.

7. Conclusion

There is clear evidence that SPS technology and TSS methodology have yielded quite some progressive results in the production of functional nanoceramic materials. Moreover, the use of modified TSS methodology in SPS equipment has shown great potential for yielding nanostructured materials with minimum risk of grain growth. However, what still remains controversial is the consistency of the functional properties and reproducibility of the methodologies used. Thus this area of study still remains highly energised for a broader enquiry. Furthermore, for most functional ceramic materials, nanostructuring has yielded enhanced material properties through various mechanisms. Although there is still room for improvement, it remains a challenge to material scientists and engineers alike to explore further and develop a deeper understanding of the mechanisms involved which may help achieve large increases in critical functional properties. Some of the highlighted problems which might have contributed to the inconsistences in functional properties include variations in the starting green densities and the likelihood of powder agglomeration at these finer sizes. This leads to inhomogeneous temperature distribution in samples and variations in sintered densities which has direct impact on material properties.

In conclusion, for practical purposes most of these materials have to satisfy certain conditions for this to become a reality: the synthesis route should be scalable, high quality and low cost, materials should have the ability to form dense compact nanostructured materials which are amenable to subsequent processing such as machining/device integration and lastly the nanostructured products should demonstrate enhanced functional properties over their micron-sized counterparts. This points to exciting scientific opportunities for continued research in order to gain more quantitative understanding to allow the design and optimisation of processes in the development of functional ceramic materials.

Author details

Wallace R. Matizamhuka
Department of Metallurgical Engineering, Vaal University of Technology, Vanderbijlpark, South Africa

*Address all correspondence to: wallace@vut.ac.za

References

[1] Anselmi-Tamburini U, Spinolo G, Maglia F, Tredici I, Holland TB, Mukherjee AK. Field assisted sintering mechanisms. In: Sintering. Berlin, Heidelberg: Springer; 2012. pp. 159-193. DOI: 10.1007/978-3-642-31009-6_8

[2] Matizamhuka WR. Spark plasma sintering (SPS)-an advanced sintering technique for structural nanocomposite materials. Journal of the Southern African Institute of Mining and Metallurgy. 2016;**116**(12):1171-1180. DOI: 10.17159/2411-9717/2016/v116n12a12

[3] Riedel R, Ionescu E, Chen I-W. Modern trends in advanced ceramics. Ceramics Science and Technology. Weinheim, Germany: Wiley-VCH; 2008. 3-38 p. ISBN: 978-3-527-31155-2

[4] Patel I. Ceramic Based Intelligent Piezoelectric Energy Harvesting Device. London, UK: INTECH Open Access; 2011. DOI: 10.5772/19189

[5] Uchino K. The development of piezoelectric materials and the new perspective. In: Advanced Piezoelectric Material. Sawstone, Cambridge, UK: Woodhead Publishing; 2017. pp. 1-92. DOI: 10.1016/B978-0-08-102135-4.00001-1

[6] Bowen LJ, French KW. Fabrication of piezoelectric ceramic/polymer composites by injection molding. In: ISAF'92: Proceedings of the Eighth IEEE International Symposium on Applications of Ferroelectrics. Greenvile, SC, USA: IEEE; 1992. pp. 160-163. DOI: 10.1109/ISAF.1992.300651

[7] Shin HY, Kim JH, Jang JS, Baek SM. FEA for fabrication process of PZT preform using cim. Journal of the Korean Ceramic Society. 2009;**46**(6):700-707. DOI: 10.4191/KCERS.2009.46.6.700

[8] Su B, Zhang D, Button TW. Embossing of ceramic micro-pillar arrays. Journal of the European Ceramic Society. 2012;**32**(12):3345-3349. DOI: 10.1016/j.jeurceramsoc.2012.04.009

[9] Bandyopadhyay A, Panda RK, Janas VF, Agarwala MK, Danforth SC, Safari A. Processing of piezocomposites by fused deposition technique. Journal of the American Ceramic Society. 1997;**80**(6):1366-1372. DOI: 10.1109/ISAF.1996.598197

[10] Han JS, Gal CW, Kim JH, Park SJ. Fabrication of high-aspect-ratio micro piezoelectric array by powder injection molding. Ceramics International. 2016;**42**(8):9475-9481. DOI: 10.1016/j.ceramint.2016.03.011

[11] Lóh NJ, Simão L, Faller CA, De Noni A Jr, Montedo ORK. A review of two-step sintering for ceramics. Ceramics International. 2016;**42**(11):12556-12572. DOI: 10.1016/j.ceramint.2016.05.065

[12] Tan Y, Zhang J, Wu Y, Wang C, Koval V, Shi B, et al. Unfolding grain size effects in barium titanate ferroelectric ceramics. Scientific Reports. 2015;**5**:9953. DOI: 10.1038/srep09953

[13] Somiya S. Handbook of Advanced Ceramics: Materials, Applications, Processing, and Properties. Waltam, USA: Academic Press; 2013. ISBN: 978-0-12-385469-8

[14] Breitwieser R, Acevedo U, Ammar S, Valenzuela R. Ferrite nanostructures consolidated by spark plasma sintering (SPS). In: Nanostructured Materials-Fabrication to Applications. Janeza, Croatia: IntechOpen; 2017. DOI: 10.5772/68017

[15] Filipović S, Pavlović VP, Obradović N, Paunović V, Maca K, Pavlović VB. The impedance analysis of sintered $MgTiO_3$ ceramics. Journal of Alloys and

Compounds. 2017;**701**:107-115. DOI: 10.1016/j.jallcom.2017.01.117

[16] Zhang X, Zhao LD. Thermoelectric materials: Energy conversion between heat and electricity. Journal of Materiomics. 2015;**1**(2):92-105. DOI: 10.10 16/j.jmat.2015.01.001

[17] Zhao LD, Dravid VP, Kanatzidis MG. The panoscopic approach to high performance thermoelectrics. Energy & Environmental Science. 2014;**7**(1): 251-268. DOI: 10.1039/C3EE43099E

[18] He J, Kanatzidis MG, Dravid VP. High performance bulk thermoelectrics via a panoscopic approach. Materials Today. 2013;**16**(5):166-176. DOI: 10.1016/j.mattod.2013.05.004

[19] Dresselhaus M. Overview of thermoelectrics for thermal to electrical energy conversion. AIP Conference Proceedings. 2013;**1519**(1):36-39. DOI: 10.1063/1.4794704

[20] Bloxam AG. Improved manufacture of electric incandescence lamp filaments from tungsten or molybdenum or an alloy thereof. GB Patent. 1906;**27**:13

[21] Tokita M. Recent and future progress on advanced ceramics sintering by Spark Plasma Sintering. Nanotechnologies in Russia. 2015;**10**(3-4):261-267. DOI: 10.1134/S1995078015020202

[22] Suarez M, Fernandez A, Menendez JL, Torrecillas R, Kessel HU, Hennicke J, et al. Challenges and opportunities for spark plasma sintering: A key technology for a new generation of materials. In: Ertu B, editor. Sintering Applications. Janeza, Croatia: InTech; 2013. pp. 319-342. DOI: 10.5772/53706. Ch. 13

[23] Guillon O, Gonzalez-Julian J, Dargatz B, Kessel T, Schierning G, Räthel J, et al. Field-assisted sintering technology/spark plasma sintering: Mechanisms, materials, and technology developments. Advanced Engineering Materials. 2014;**16**:830-848. DOI: https://doi.org/10.1002/adem.201300409

[24] Chen IW, Wang XH. Sintering dense nanocrystalline ceramics without final-stage grain growth. Nature. 2000;**404**(6774):168, 35004548

[25] Chu MY, De Jonghe LC, Lin MK, Lin FJ. Precoarsening to improve microstructure and sintering of powder compacts. Journal of the American Ceramic Society. 1991;**74**(11):2902-2911. DOI: 10.1111/j.1151-2916.1991.tb06861

[26] Wang XH, Chen PL, Chen IW. Two-step sintering of ceramics with constant grain-size, I. Y_2O_3. Journal of the American Ceramic Society. 2006;**89**(2):431-437. DOI: 10.1111/j.1551-2916.2005.00763.x

[27] Sutharsini U, Thanihaichelvan M, Singh R. Two-step sintering of ceramics. Sintering of Functional Materials. 2017:1-20. DOI: 10.5772/68083

[28] Ferreira GBP, da Silva JF Jr, do Nascimento RM, Gomes UU, Martinelli AE. Two-step sintering applied to ceramics. In: Sintering of Ceramics-New Emerging Techniques. Janeza, Croatia: InTech; 2012. ISBN 978-953-51-0017-1

[29] Mazaheri M, Zahedi AM, Sadrnezhaad SK. Two-step sintering of nanocrystalline ZnO compacts: Effect of temperature on densification and grain growth. Journal of the American Ceramic Society. 2008;**91**(1):56-63. DOI: 10.1111/j.1551-2916.2007.02029.x

[30] Mazaheri M, Simchi A, Golestani-Fard F. Densification and grain growth of nanocrystalline 3Y-TZP during two-step sintering. Journal of the European Ceramic Society. 2008;**28**(15):2933-2939. DOI: 10.1016/j.jeurceramsoc.2008.04.030

[31] Kim HT, Han YH. Sintering of nanocrystalline $BaTiO_3$. Ceramics International. 2004;**30**(7):1719-1723. DOI: 10.1016/j.ceramint.2003.12.141

[32] Wang JC, Zheng P, Yin RQ, Zheng LM, Du J, Zheng L, et al. Different piezoelectric grain size effects in $BaTiO_3$ ceramics. Ceramics International. 2015;**41**(10):14165-14171. DOI: 10.1016/j.ceramint.2015.07.039

[33] Karaki T, Yan K, Adachi M. Barium titanate piezoelectric ceramics manufactured by two-step sintering. Japanese Journal of Applied Physics. 2007;**46**(10S):7035. DOI: 10.1143/JJAP.46.7035/meta

[34] Wu MW. Two-step sintering of aluminum-doped zinc oxide sputtering target by using a submicrometer zinc oxide powder. Ceramics International. 2012;**38**(8):6229-6234. DOI: 10.1016/j.ceramint.2012.04.076

[35] Mazaheri M, Zahedi AM, Haghighatzadeh M, Sadrnezhaad SK. Sintering of titania nanoceramic: Densification and grain growth. Ceramics International. 2009;**35**(2):685-691. DOI: 10.1016/j.ceramint.2008.02.005

[36] Suárez G, Sakka Y, Suzuki TS, Uchikoshi T, Zhu X, Aglietti EF. Effect of starting powders on the sintering of nanostructured ZrO_2 ceramics by colloidal processing. Science and Technology of Advanced Materials. 2009;**10**(2):025004. DOI: 10.1088/1468-6996/10/2/025004

[37] Xiong Y, Hu J, Shen Z. Dynamic pore coalescence in nanoceramic consolidated by two-step sintering procedure. Journal of the European Ceramic Society. 2013;**33**(11):2087-2092. DOI: 10.1016/j.jeurceramsoc.2013.03.015

[38] Schwarz S, Guillon O. Two step sintering of cubic yttria stabilized zirconia using field assisted sintering technique/spark plasma sintering. Journal of the European Ceramic Society. 2013;**33**(4):637-641. DOI: 10.1016/j.jeurceramsoc.2012.10.002

[39] Papynov EK, Tkachenko IAE, Portnyagin AS, Modin EB, Avramenko VA. Fabrication of magnetic ceramic materials based on nanostructured hematite powder by spark plasma sintering. Carbon (C). 2006;**4**(243):2-5. ISSN 1819-6608

[40] Saravanan P, Hsu JH, Sivaprahasam D, Kamat SV. Structural and magnetic properties of γ-Fe_2O_3 nanostructured compacts processed by spark plasma sintering. Journal of Magnetism and Magnetic Materials. 2013;**346**:175-177. DOI: 10.1016/j.jmmm.2013.07.023

[41] Gaudisson T, Vázquez-Victorio G, Bañobre-López M, Nowak S, Rivas J, Ammar S, et al. The Verwey transition in nanostructured magnetite produced by a combination of chimie douce and spark plasma sintering. Journal of Applied Physics. 2014;**115**(17):17E117. DOI: 10.1063/1.4863164

[42] Gaudisson T, Acevedo U, Nowak S, Yaacoub N, Greneche JM, Ammar S, et al. Combining soft chemistry and spark plasma sintering to produce highly dense and finely grained soft ferrimagnetic $Y_3Fe_5O_{12}$ (YIG) ceramics. Journal of the American Ceramic Society. 2013;**96**(10):3094-3099. DOI: 10.1111/jace.12452

[43] Boda SK, Thrivikraman G, Panigrahy B, Sarma DD, Basu B. Competing roles of substrate composition, microstructure, and sustained strontium release in directing osteogenic differentiation of hMSCs. ACS Applied Materials & Interfaces. 2016;**9**(23):19389-19408. DOI: 10.1021/acsami.6b08694

[44] Ovtar S, Le Gallet S, Minier L, Millot N, Lisjak D. Control of barium ferrite decomposition during

spark plasma sintering: Towards nanostructured samples with anisotropic magnetic properties. Journal of the European Ceramic Society. 2014;**34**(2):337-346. DOI: https://doi.org/10.1016/j.jeurceramsoc.2013.07.027

[45] Hungria T, Galy J, Castro A. Spark plasma sintering as a useful technique to the nanostructuration of piezo-ferroelectric materials. Advanced Engineering Materials. 2009;**11**(8): 615-631. DOI: https://doi.org/10.1002/adem.200900052

[46] Arlt G, Hennings D, De With G. Dielectric properties of fine-grained barium titanate ceramics. Journal of Applied Physics. 1985;**58**(4):1619-1625. DOI: 10.1063/1.336051

[47] Takeuchi T, Betourne E, Tabuchi M, Kageyama H, Kobayashi Y, Coats A, et al. Dielectric properties of spark-plasma-sintered $BaTiO_3$. Journal of Materials Science. 1999;**34**(5):917-924. DOI: 10.1016/j.msec.2016.02.027

[48] Wang K, Li JF, Zhou JJ. High normalized strain obtained in Li-modified (K, Na) NbO_3 lead-free piezoceramics. Applied Physics Express. 2011;**4**(6):061501. DOI: 10.1143/APEX.4.061501

[49] Zhang MH, Wang K, Du YJ, Dai G, Sun W, Li G, et al. High and temperature-insensitive piezoelectric strain in alkali niobate lead-free perovskite. Journal of the American Chemical Society. 2017;**139**(10): 3889-3895. DOI: 10.1021/jacs.7b00520

[50] Chen C, Liang R, Zhou Z, Zhang W, Dong X. Enhanced bipolar fatigue resistance in PMN-PZT ceramics prepared by spark plasma sintering. Ceramics International. 2018;**44**(4):3563-3570. DOI: 10.1016/j.ceramint.2017.11.051

[51] Han B, Zhao C, Zhu Z-X, Chen X, Han Y, Duan H, et al. Temperature-insensitive piezoelectric performance in Pb $(Zr_{0.52}Ti_{0.42}Sn_{0.02}Nb_{0.04})O_3$ ceramics prepared by spark plasma sintering. ACS Applied Materials & Interfaces. 2017;**9**(39):34078-34084. DOI: 10.1021/acsami.7b09825

[52] Shen ZY, Li JF, Wang K, Xu S, Jiang W, Deng Q. Electrical and mechanical properties of fine-grained Li/Ta-modified (Na, K) NbO_3-based piezoceramics prepared by spark plasma sintering. Journal of the American Ceramic Society. 2010;**93**(5):1378-1383. DOI: 10.1111/j.1551-2916.2009.03542.x

[53] Zhang QM, Wang H, Kim N, Cross LE. Direct evaluation of domain-wall and intrinsic contributions to the dielectric and piezoelectric response and their temperature dependence on lead zirconate-titanate ceramics. Journal of Applied Physics. 1994;**75**(1):454-459. DOI: 10.1063/1.355874

[54] Ochoa DA, Esteves G, Iamsasri T, Rubio-Marcos F, Fernández JF, García JE, et al. Extensive domain wall contribution to strain in a (K, Na) NbO_3-based lead-free piezoceramics quantified from high energy X-ray diffraction. Journal of the European Ceramic Society. 2016;**36**(10):2489-2494. DOI: 10.1016/j.jeurceramsoc.2016.03.022

[55] Chen ZG, Han G, Yang L, Cheng L, Zou J. Nanostructured thermoelectric materials: Current research and future challenge. Progress in Natural Science: Materials International. 2012;**22**(6): 535-549. DOI: 10.1016/j.pnsc.2012.11.011

[56] Tritt TM, Böttner H, Chen L. Thermoelectrics: Direct solar thermal energy conversion. MRS Bulletin. 2008;**33**(4):366-368. DOI: 10.1557/mrs2008.73

[57] Yu BL, Qi Q, Tang XF, Zhang QJ. Effect of grain size on thermoelectric

properties of $CoSb_3$ compound. Acta Physica Sinica. 2005;**54**:5763-5768

[58] Zhao XY, Shi X, Chen LD, Zhang WQ, Bai SQ, Pei YZ, et al. Synthesis of $Yb_yCo_4Sb_{12}/Yb_2O_3$ composites and their thermoelectric properties. Applied Physics Letters. 2006;**89**(9):092121. DOI: 10.1063/1.2345249

[59] Lee Y, Lo SH, Androulakis J, Wu CI, Zhao LD, Chung DY, et al. High-performance tellurium-free thermoelectrics: All-scale hierarchical structuring of p-Type PbSe–MSe Systems (M= Ca, Sr, Ba). Journal of the American Chemical Society. 2013;**135**(13):5152-5160. DOI: 10.1021/ja400069s

[60] Zhao LD, He J, Wu CI, Hogan TP, Zhou X, Uher C, et al. Thermoelectrics with earth abundant elements: High performance p-type PbS nanostructured with SrS and CaS. Journal of the American Chemical Society. 2012;**134**(18):7902-7912. DOI: 10.1021/ja301772w

[61] Zhao LD, Zhang BP, Li JF, Zhou M, Liu WS, Liu J. Thermoelectric and mechanical properties of nano-SiC-dispersed Bi_2Te_3 fabricated by mechanical alloying and spark plasma sintering. Journal of Alloys and Compounds. 2008;**455**(1-2):259-264. DOI: 10.1016/j.jallcom.2007.01.015

[62] Xiong Z, Chen X, Zhao X, Bai S, Huang X, Chen L. Effects of nano-TiO_2 dispersion on the thermoelectric properties of filled-skutterudite $Ba_{0.22}Co_4Sb_{12}$. Solid State Sciences. 2009;**11**(9):1612-1616. DOI: 10.1016/j.solidstatesciences.2009.06.007

[63] Noguchi T. Powder processing of thermoelectric materials-focusing on SiGe with new sintering technique. In: XVI ICT'97. Proceedings ICT'97. 16th International Conference on Thermoelectrics. August 1997. pp. 207-214. DOI: 10.1109/ICT.1997.667082

[64] Möchel A, Sergueev I, Wille HC, Juranyi F, Schober H, Schweika W, et al. Lattice dynamics in the thermoelectric Zintl compound $Yb_{14}MnSb_{11}$. Physical Review B;**84**(18):184303. DOI: 10.1103/PhysRevB; 2011.84.184303

[65] Yu C, Chen Y, Xie H, Snyder GJ, Fu C, Xu J, et al. Improved thermoelectric properties in Lu-doped $Yb_{14}MnSb_{11}$ Zintl compounds. Applied Physics Express. 2012;**5**(3):031801. DOI: 10.1143/APEX.5.031801

[66] Wang S, Yang J, Wu L, Wei P, Yang J, Zhang W, et al. Anisotropic multicenter bonding and high thermoelectric performance in electron-poor CdSb. Chemistry of Materials. 2015;**27**(3):1071-1081. DOI: 10.1021/cm504398d

[67] Jiang Q, Yang J, Liu Y, He H. Microstructure tailoring in nanostructured thermoelectric materials. Journal of Advanced Dielectrics. 2016;**6**(01):1630002. DOI: 10.1142/S2010135X16300024

Chapter 8

Rapid Physical Models: A New Phase in Industrial Design

Peer M. Sathikh

Abstract

Rapid prototyping, especially in the form of 3D printing, has pervaded over key aspects of design engineering since the start of this millennium. Today, rapid physical model making has applications in engineering, architecture, design, and fine art. While 3D printing today is mostly about prototyping of design as a precursor to production, not many have studied the use of 3D for industrial design in detail. With core responsibilities for three important nodes of user experience, namely function, human factors (ergonomics) and the aesthetics and emotion, 3D printing has been playing a major role in the process of industrial design. This chapter elucidates this through examples leading the reader to think about the future practice of rapid physical model making in industrial design. The chapter concludes by mentioning future scenarios that industrial design may take with constant innovations in 3D printing.

Keywords: Industrial design, form and function, rapid physical models, 3D printing

1. Introduction

Industrial design as a profession started around 100 years ago, stemming from the fact that the end of the eighteenth century saw two developments which required an entirely different approach to products. Firstly, products were no longer ‘artefacts’ but products themselves. Secondly, the emergence of the concept of mass consumption which sounded the end of ‘industrial art’ which was more interested in the decorative elements of the product rather than the appropriate aesthetics for the product which considers both the function and form. According to the Conran Directory of Design [1], it was in 1910, in Germany, when an architect named **Peter Behrens** took complete charge of the all aspects of the appearance of an industrial corporation for AEG, designing everything from products such as table fans and electric kettles to posters to the interior design of the building itself (**Figure 1**)[1] that the first signs of industrial design as a profession.

With new materials and production methods emerging at the end of the 19th century, the thought of formalising design education programme, separate from arts and handicrafts, began to take shape in Europe. It is with the founding of the **Bauhaus** in Weimar, Germany in 1919 by **Walter Gropius**, that design as a profession for the modern world became recognised. Bauhaus in its short life spanned a design philosophy which still has influence today. On the other side of the Atlantic,

[1] Source–left: http://www.sothebys.com/en/auctions/ecatalogue/2007/deutscher-werkbund-to-bauhaus-an-important-collection-of-german-design-n08459/lot.46.html
Source–right: https://www.ft.com/content/a0d0b9b8-4245-11e8-97ce-ea0c2bf34a0b

Figure 1.
Industrial design for AEG by Peter Behrens.

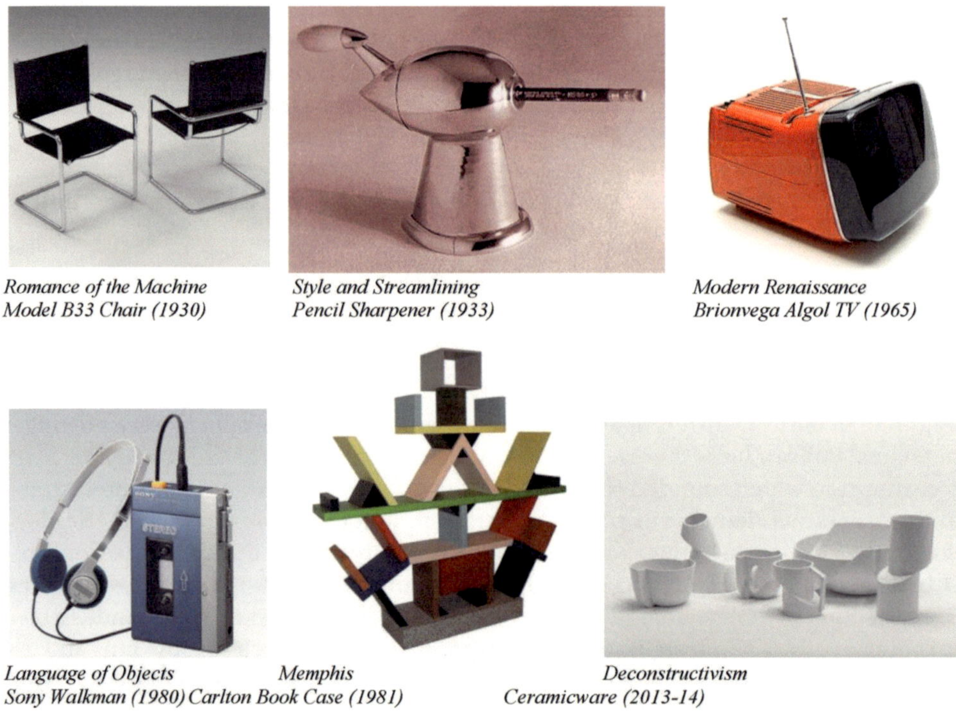

Figure 2.
Industrial design through different phases/movements.

several key players such as Norman Bel Geddes, Walter Dorwin Teague, Henry Dreyfuss, Harold Van Doren and **Raymond Lowey**, each from different background started plying their trade as consultant designers around the same time.

Industrial design, since then, has traversed through several phases, more or less following the progress of technology in varied fields, including materials and manufacturing, computer science & engineering and information technology to where it is today. According to Stephen Bayley, the editor of the Conran Directory of

Figure 3.
Lowey's redesign of Gestetner duplicator; first use of clay to model the form (source: https://collections.vam.ac.uk/item/O322014/gestetner-duplicator-duplicator-loewy-raymond-fernand/#).

Design [1], industrial design has passed through mass consumption (beginning of consumer age), the modern movement (the romance of the machine), the style era, modern renaissance, the language of objects (symbolism and consumer psychology) till the 1980s. Since then, the postmodern era has seen some radical design movements such as the Memphis and Deconstructivism. Significant and iconic products of each phase are seen in **Figure 2**.

The biggest effect of industrial design, right from its earliest beginnings, has been the use of physical models during the design process. Stephen Bayley [1] writes that, 'In his first major job, in 1929 for the English reprographic machine manufacturer Sigmund Gestetner (**Figure 3**), he fused the spirit of the times with 50 lbs of clay and made the first piece of office equipment to rely on streamlining'. Why did Lowey use clay? He himself explains that, 'And because Gestetner needed the design so quickly, there was no way to work in steel. I kept as close to the skeleton as possible to be efficient' [2]. Since then, physical models, to check not only the aesthetics, but also the ergonomics, function and dimensional fit of products, have been used extensively in industrial design.

2. Physical models in industrial design

Industrial designers have been using physical models in the design process in many ways. What started off as a means to portray form and aesthetics in a three-dimensional format, physical models have evolved to be used for many intentions and purposes during the different stages of design. Towards the mid-1980s, the design process was more stabilised with distinctive and accepted stages as ideas progressed through as shown by the simplified diagram in **Figure 4**.

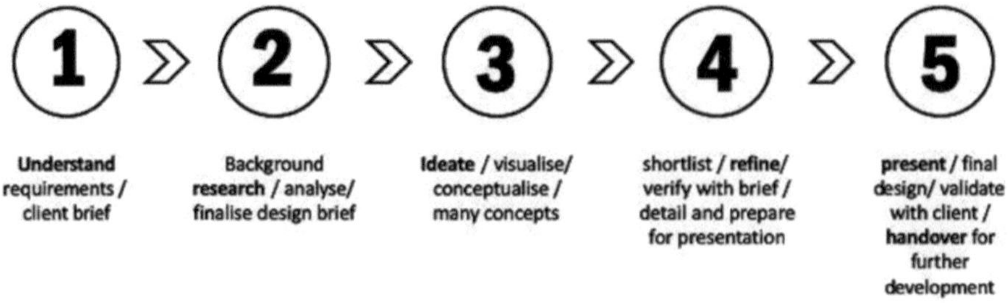

Figure 4.
Industrial design process.

Figure 5.
Study models (source: http://whiteboardps.com/index.php/fwp_portfolio/acco-swingline-gbc-fusion-laminators/).

2.1 Non-functional models

Referring to **Figure 4**, it is during stages 3 and 4 that physical models play a very important part in industrial design. During Stage 2, designers rely on sketching as an ideation tool. Traditionally, it was common (and still is) to produce concept sketches using pencil/ball pen/felt tip pen on paper generation concepts in freehand perspectives. While perspective drawings convey the form and shape of the product being conceptualised, not all people can interpret the sketches and understand the nuances of the design concept being represented in those sketches.

For that reason, industrial designers resort to fabricating study models in foam, clay, plaster or any other material to supplement the concept sketches (**Figure 5**). Many a times these study models are made in full scale for the client and/or potential users to get a feel of the size, form and fit as well as the design details.

When products such as a car is being designed, study models take on a new meaning. It is a common practice to sculpt full scale models of the exterior using clay. This allows for not only verification of the form and the subtle carvings on the surface, it also allows for making changes to these surface curves and details which can then be captured back into detail design. **Figure 6** shows the clay modelling facility of a car manufacturer.

According to the Modelling Manager at Ford,[2] a full-size clay model of a vehicle allows the designers and engineers to spot potential issues in both the interior and exterior of the vehicle which are not apparent on digital or small-scale models. With the advance reverse engineering technology available designers model the changes that necessary by hand and scan those changes back into the computer in order to capture it and integrate this into the main 3D data to integrate into the final design.

[2] https://social.ford.com/en_US/story/ford-community/automotive-news/we-reveal-a-ford-automotive-design-secret.html

Figure 6.
Clay modelling in full scale (source: https://i.pinimg.com/originals/40/fa/22/40fa22a8d8120cceed06deeb030 2db12.jpg).

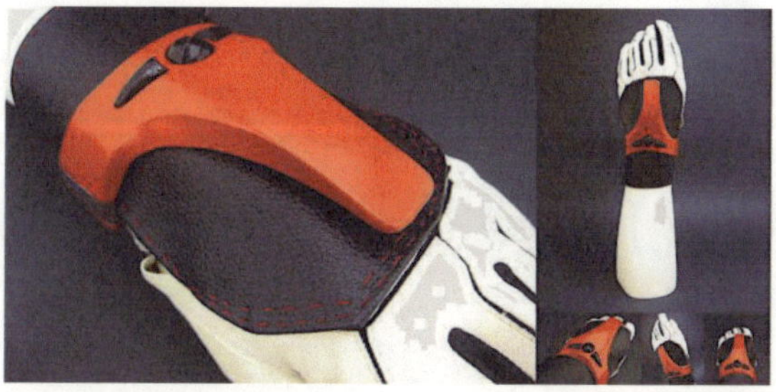

Figure 7.
Full scale mock up model.

Figure 8.
Mock up model of electric scooter (source: https://www.designideas.pics/porter/).

During Stage 4 of the design process shown in **Figure 4**, many times, full scale appearance model is presented to the client/audience which shows the exterior details in full including colour, texture and graphics as seen in **Figure 7**. It is a common practice to present more than one concept, hence more than one full scale mock up models are presented. In the case of vehicle/car design, scale models are presented as shown in **Figure 8**.

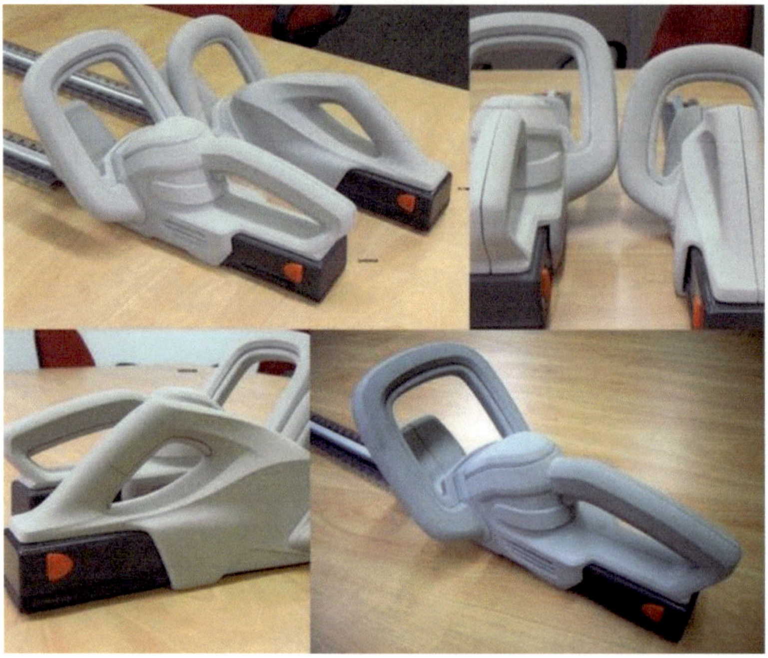

Figure 9.
Foam models to check form and ergonomic fits (source: https://www.pinterest.com/pin/289497082275373127/).

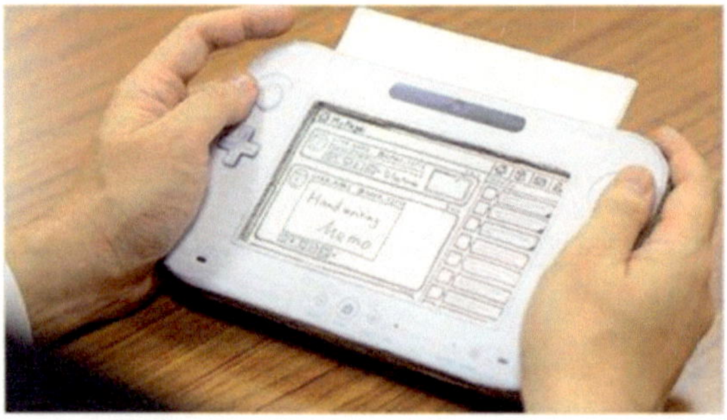

Figure 10.
Low fidelity mock up (source: https://engineeringproductdesign.com/).

Physical models in industrial design play a bigger role in design decision making than just iterating design/aesthetic variations including partially checking the function and fit of the product being designed.

2.2 Semi-functional models

At the early stages of the design process (Stage 2 in **Figure 4**), physical models play an important role in ascertaining human factors/ergonomics aspects of the product, besides presenting the form and fit. An example is shown in **Figure 9** where the ergonomics of a hand held power tool is important to the success of the product. Variations in the hand hold areas allow potential users to test/play act and give feedback on the right design for such areas.

In designing products with electronic displays and interface, it is a common practice to embed such components into the full scale mock-ups, many a times with

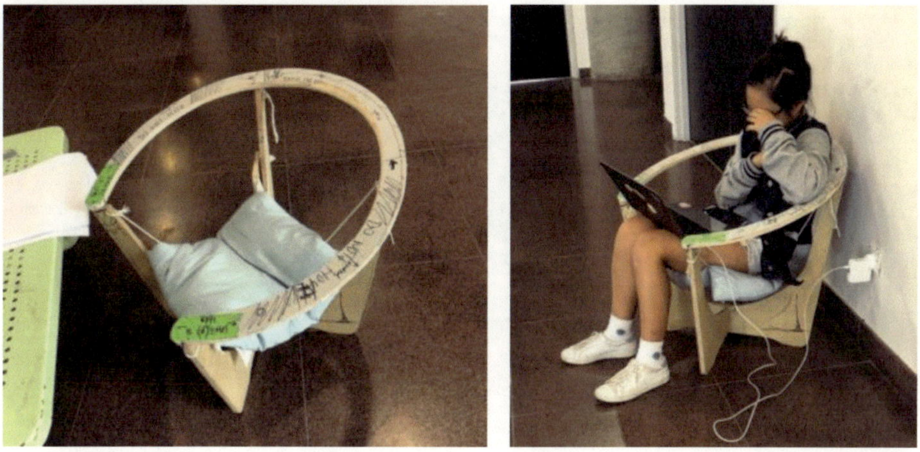

Figure 11.
Quick furniture model.

rudimentary connections, to enable users to test the first level comfort in user experience. Such a mock up model as shown in **Figure 10** is also termed as' low fidelity' mock up in the industry. A low fidelity mock up allows the design team to verify the functional concept by allowing users to 'playact' the way they probably will use the real product. Designers can quickly make changes to the model based on feedback and make modifications. The almost real-time iteration of the design, allowing for variations in functionality make low-fidelity models an important part of industrial design in the age of interaction design.

Furniture designers use quick physical models to verify the concept in terms of aesthetics, manufacturing process and human fits. Such a quick model made by a student of furniture design is shown in **Figure 11**.

3. Classifying physical models

Physical models in product development in general, and industrial design in specific, may be classified in several ways. It is best to understand these classifications in order to understand the impact of digitalization and rapid model making has on industrial design.

Broek et al. [3] classify and exactly describe physical models according to usage and type as:

1. *Visualisation*: models are used for presentations and shape (details). They can support reasoning about shape geometry, curvature and accuracy, texture, colour, finishing, and graphics. Shapes become tangible, local curvature and product appearance can be judged.

2. *Functionality testing*: depending on the tested functions, the model representation is not too precise at those regions where no testing is performed. However, the degrees of freedom for optimal testing must be guaranteed, and testing regions, e.g., ergonomic verification, must be represented accurately.

3. *Physical testing*: a materialised model must be fabricated consisting of the same material of the final product. Accuracy and exclusion of strength variations related to the fabrication technology are important issues.

Soft Model	Hard Model	Presentation Model	Prototype
• rough modelling • use to assess the overall size, proportion, and shape of many proposed concept. • fast evaluation of basic sizes and proportions • reshaped and refined by hand to explore and improve its tactile quality	• technically non-functional yet are close replicas of the final design • made from wood, dense foam, plastic, or metal are painted and textured • have some "working" features such as button that push or sliders that move	• model that constructed and matched from CAD data • complete model and fully detailed composition of the product • component of this model will be simplified or neglected due to cost or time shortages	• high-quality model or functioning product that is produce to realize a design solution. • would be tested and evaluated before the product is considered for production.

Table 1.
Classification of physical models [4].

4. *Marketing*: a marketing model or presentation model will express the added design value of the product to outsiders of the design process. The finishing quality and being a look-alike of the final product are crucial for this type of models.

5. *Proof-of-concept*: a very detailed model made in the final stage of design to qualify the product design against the requirements.

6. *Editing*: editable models are assembled or composed models and, when needed, decomposed again and rebuild with different (shape) components to create an adapted version of the same model.

7. *Communication*: a communication model is applicable for communication with the inside of the design process or for explanation to the related authorities to provide them with a better understanding what is going on in the design process.

8. *Process*: a process model is a kind of proto-model or protoshape like a CAD design or a physical model, which is treated in a reverse engineering way. In those models the progress of a design is captured, and the shape of a model can be change manually [3].

Of the classification by Broek et al. [3], classic industrial design is interested in *visualisation, functionality-testing, marketing,* and *proof-of-concept* models. Isa and Liem [4] state that there are, '... very limited classifications which clearly explained the actual characteristics and functions of each physical models in the design process' and that the lack of classifications makes it, '... harder for the designer to understand the true potential of physical models in various fields'. Isa and Liem [4] give a first level classification as shown in **Table 1**.

Isa and Liem [4] have also elaborated on the classification from Broek [3] which is shown in **Table 2**.[3] What **Tables 1** and **2** show are where the practice of physical model making in the design process has arrived at in the first part of the 21st century. This has been made possible through several key factors along the way. It

[3] Isa and Liem [4] have included Technology and have excluded Editing in the classification shown in **Table 2**. The references cited in **Table 2** refers to the source article [4].

Usage \ Types	Soft Model [Ulrich and Eppinger 2012]	Hard Model [Ulrich and Eppinger 2012]	Presentation Model [Kimoji and Tano 1991]	Prototype [Ulrich and Eppinger 2012]
Visualization [Broek et al. 2009]	• Visualization tool for early insights [Masctelli 2000]	• Support about shape, function, geometry, colour [Broek et al. 2009]	• Represent outer appearance of the design, visualisation of total design [Broek et al. 2009]	• CAD , detail design stage , very detailed model [Broek et al. 2009]
Functionality testing *(Ergonomic Testing)* [Broek et al. 2009]	• Cannot be tested with actual usage, not functional [Broek et al. 2009]	• Can be tested with actual size but with not full function criteria [Broek et al. 2009]	• Some part of the design can be fully tested. [Broek et al. 2009]	• correct interpretation of ergonomic data or of good practice in the measurement of individual subjects. [Broek et al. 2009]
Physical testing [Broek et al. 2009]	• Depending on the tested function [Broek et al. 2009] • Principal testing [Sæter et al. 2012]	• Depending on the tested function [Broek et al. 2009] • Form and shape testing [Sæter et al. 2012]	• Can stimulate certain behaviour like strength and stiffness. [Broek et al. 2009]	• Final trade-off of performances [Masctelli 2000] • Fully functional model [Sæter et al. 2012]
Marketing [Broek et al. 2009]	• product appearance can be judged [Broek et al. 2009]	• Incorporate early feedback from customers [Masctelli 2000]	• Express the added design value of product to outsiders [Broek et al. 2009]	• Results in higher user satisfaction [Broek et al. 2009]
Proof of concept [Broek et al. 2009]	• Initial early stage model [Ulrich and Eppinger 2012] • Basic model [Sæter et al. 2012]	• Semi detail model [Ulrich and Eppinger 2012] • Complex shape [Sæter et al. 2012]	• Completely finished :color, gloss, texture etc. [Broek et al. 2009] • exact feel and look [Sæter et al. 2012]	• A very detail model in the final stage of design to qualify the product design against requirements. [Broek et al. 2009]
Editing [Broek et al. 2009]	• decomposed again, rebuild with different shape [Broek et al. 2009] • lots of modification [Sæter et al. 2012]	• When needed decomposed again and rebuild with different material and adjustment of the shape. [Broek et al. 2009]	• Editable models are assembled or final composed model [Broek et al. 2009]	• Not editable and will lead to higher cost [Broek et al. 2009] • Very minor adjustments [Sæter et al. 2012]
Technology [Broek et al. 2009]	• Very inexpensive and quick [Lafon and Mackav management and customers [Masctelli 2000]	• Not complex technology and manual handmade [Broek et al. 2009] management [Masctelli 2000]	• Expose designers to potential future system enhancements [Broek et al. 2009] developers and end users. [Lafon and Mackay 2000]	• Complexity technology of manufacturing, complex in terms of the same as the prototype [Broek et al. 2009]

Table 2.
Classification of physical models according to usage [4].

is important to understand what has changed from 100 years ago as summarised in **Table 3** showing the progress based on the materials and the method of fabrication/manufacture of the physical models. It can be seen from **Table 3** that, as the method of fabrication of physical models progressed from hand fabricated models to automated model making, the accuracy of the dimensions and the ability to

	Hand Fabrication	Basic Machine Based	NC Machine Tools	CNC Machine Tools	High Speed CNC	Laser Cutter	1st Generation 3D Printing	2nd Generation 3D Printing	3rd Generation 3D Printing
Clay	●				●				
Plaster	●	●							
Foam	●	●	●	●					
Polystyrene		●	●	●					
ABS		●	●	●	●				
Acrylic		●	●	●		●			
MDF Board		●	●	●		●			
Chemical Wood		●	●	●	●				
Resin							●		
Filament							●	●	●
	Traditional				**Rapid Models**				

Table 3.
Progress of physical model making for industrial design.

realise models of complex/sophisticated form is increased. This shift from simple methods to complex process in model making has been accompanied by technological advancement, mainly in computer aided design (CAD) and computer numerical control (CNC) which are the keys to today's model making techniques such as high speed CNC, laser cutting and 3D printing. The availability of accurate 3D data produced by CAD software enables reliable CNC. According to Mike Lynch,[4] Founder and President of CNC Concepts Inc., CNC offers three distinct benefits, the first being the **reduced skill level** due to improved automation. The second benefit is the **consistency and accuracy** of the parts that are produced and third benefit is the **flexibility** to change to many different parts or models.

During the starting period of model making during the last century, the emphasis laid on the skill of the model maker. Those who took up this profession were craftsmen or, many a times, the designers. The dimensional accuracy of the model.

as well as the shape, form and finishes depended on the skill level of the model maker. With the advent of 'rapid' methods this skill was embedded in the machine and method itself and the model maker has more of a technician's role in the model making process. How did this change happen? This is best explained in the next section on computer aided design (CAD).

3.1 Enter CAD

The most significant progress in product design and development occurred with the advent of computer aided design (CAD) in the 1960s. CAD as an idea and working prototype was derived from the idea of CNC (which was developed by Dr. Patrick J. Hanratty in 1957) and first developed by Ivan Sutherland at MIT as SKETCHPAD which showed the capabilities of computers with display in technical drawing. The full potential of CAD as a three-dimensional development tool was realised through

[4] https://www.mmsonline.com/articles/key-cnc-concept-1the-fundamentals-of-cnc

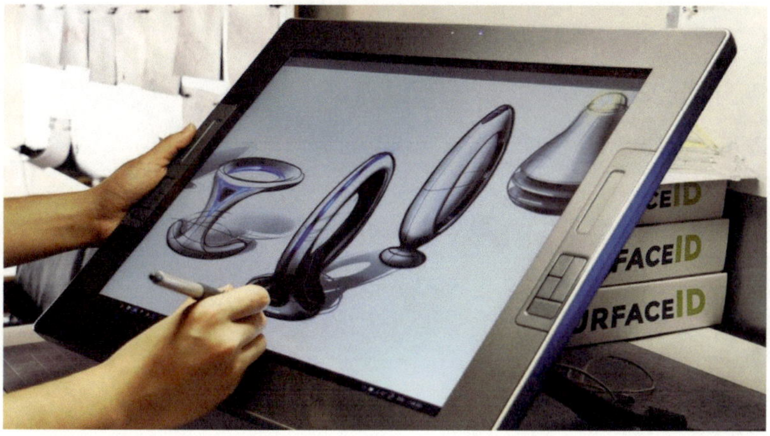

Figure 12.
CAD sketching (source: surfaced.com).

software such as Pro/Engineer, UniGraphics, CATIA in the 1980s. With CAD, industrial designers were able to design in 3D and define the details on the surface and the various features within the 3D environment, which then becomes 3D data that is utilised by any CNC controlled system to machine or manufacture the design with a high level of fidelity in terms of dimensions and details. Today industrial designers use very sophisticated software such as Autodesk 360 and Rhinoceros, which allows for not only 3D data transfer for model making, but also for 'photo-realistic' renderings and to transfer the model to engineers for detail development. **Figure 12** shows the high level of flexibility that CAD offers including visualisation in sketch form which then could be automatically converted into 3D data.

Having 3D data that is transferrable with high reliability of the design concepts is the first step towards the realisation of rapid physical models. Many formats for transferring reliable 3D data have been developed, the most common ones being IGES, STEP, STL and OBJ. In the field of rapid model making, IGES and STEP are predominantly used in high speed CNC while STL and OBJ formats are the most reliable for 3D printing.[5] IGES is the earliest format and is still a popular, though more for 2D graphics and object rather than for 3D format. STEP is perhaps the first true 3D file convertor which relates to ISO 10303 and is widely used to transfer 3D data created by different CAD software platforms as well as transfer data to CNC programmers. STL is pure 3D information on geometry and shapes and does not hold any information colour, textures, etc. OBJ file format stores both form (geometry and shape) data as well as colour and texture information and is very useful in the latest multicolour 3D printers with high resolution capabilities.

4. Rapid physical models

Three distinct advances in machining technology has paved the way for rapid physical models in the last 25 years or so. At first, emergence of **high-speed CNC** milling machines in various sizes allowed model makers to fully utilise its capabilities for model making for industrial design.

Second is the emergence of two-dimensional computer-controlled **laser cutters**, which operate on the same drafting principle of a graphic plotter. This allowed for quick machining of 2D shapes in both opaque and transparent materials. It has been

[5] https://www.cadcrowd.com/blog/top-file-formats-for-sharing-3d-and-2d-cad-designs/

Figure 13.
Concept irons from students.

Figure 14.
3D printed spectacle frames (source: 3ders.org).

said that the famous Silicon Valley in California, USA, has overplayed the contribution of laser cutters in the development of hi-tech products.

Third, and the most influential technology in the development of rapid model making is introduction of additive manufacturing, which is also known as **3D printing**. 3D printing is not one technology but many that has been developed by several companies/corporation based on the principle that a model can be built, layer by layer, using materials that are deposited or extruded and cured to solid state as the layers are added. Looking at the history of 3D printing the first technology to emerge around 1984 was **Stereolithography** followed by Selective Laser Sintering (**SLS**) in 1988 which was then followed by Fusion Deposition Modelling (**FDM**). Several companies manufacture 3D printers in what is now a competitive market which cater to institutions and professional companies/outfit that could purchase and maintain a high-end rapid model making equipment. Since 2010, table top 3D printers that are much lower in cost and easy to maintain have become popular allowing not only designers, but craftspeople and hobby enthusiasts to purchase them.

4.1 Advantage of rapid physical modelling

The obvious advantage of rapid physical model making is in the saving of time and the human centric energy required to craft the models. This is especially so in the case of 3D printing where the need for assembling several fabricated parts to make a whole (model) is eliminated. This ability is more useful when early visualisation models that do not require finishes are made for form, fit and ergonomic testing.

DOI: http://dx.doi.org/10.5772/intechopen.88788

Desktop 3D printers and the economics it has brought also proves advantageous to students of industrial design, allowing them to make several iterations and variations of concepts within a short period of time allowing them the opportunity to explore options and discuss the pros and cons with the instructors before deciding on a final design. This option was mostly available as two-dimensional sketch and render exploration in the pre 3D printing era. **Figure 13** shows the result of two such student exercise where the final designs were 3D printed and finished with colour and details after preliminary models were printed for exploration.

Such an advantage of design exploration is also possible by professional designers and design companies, who have gone beyond the polished renderings to study and refine the overall design of the products due to the affordance that 3D printing offers. **Figure 14** shows 3D printed exploration and production spectacle frames.

The end result of such explorations seems to point to a new phase in industrial design where the traditional design process has been disrupted by CAD and rapid physical models. New concepts such as collaborative design has brought in real time design and development through networked connectivity and the Internet. As the profession of industrial design has done in the last century, it is evolving to meet this change.

5. The new phase in industrial design

With CAD, the Internet, networking, cloud computing, and rapid physical model making (as well as prototyping) engulfing the day-to-day activities of an industrial designer, what then could be the new phase of industrial design. These are some of the aspects of design and process that has changed:

a. **Change in the process flow:** The traditional approach to design process depict in **Figure 4** gives way to more multi-disciplinary process as shown in **Figure 15**. With 3D data being created, refined and evolved at the early stage of industrial design (Stage 3), first level development engineers can start working on the preliminary part of the development supported by rapid physical models that could be 'shelled' to leave void space. What was already possible with the introduction of CAD becomes much more concrete with the aid of such models that serve as low fidelity engineering prototypes.

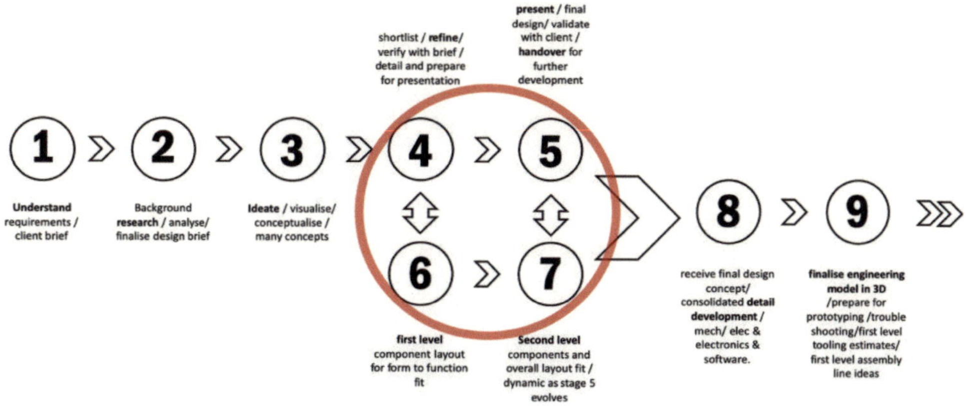

Figure 15.
Revised design process.

b. **Verification of industrial design details:** Photorealistic renderings are 2D depiction, either on the screen or on printed copies. This does not allow designers to look and feel the nuanced details and curves that form the surface of the product being designed. Rapid physical models allow for making several concepts for the same details and allow for study and discussion before a decision is made. In many cases another set of rapid physical models are built after modification for further checking. This is very true in the automotive industry where the body surfaces have subtle and nuanced curves and details. Waiting for a full-scale clay model may be too late and opportunities to make several variations get lost. **Figure 16** shows a high-speed CNC machining of automotive body in progress.

c. **Rapid physical models made simultaneously:** Since rapid physical models are made from a 3D data base created by industrial designers (and development engineers), it is possible to send the data across the world to any part of the world that has rapid model making facilities. This would allow for designers across the different places to inspect the model simultaneously and initiate variations that may be needed for the product due to various reasons, including cultural, environmental, statutory requirements or availability of components, etc.

d. **Localised manufacture:** The speed at which 3D printing technology is developing, allows for localised manufacture of products where the quantity is low, suitable for small batch manufacture. This will be a scenario where, for instance, medical support team has to build basic equipment and furniture, and perhaps shelters. Low cost, desktop 3D powered by batteries could be a solution for fabricating them to suit the environment and the environment. This idea could be extended to future manned missions to the Moon and Mars where industrial design and development work can be done on Earth and the data sent remotely for the mission personnel to manufacture and commission on sight.

e. **Small quantity and customised manufacture:** This is already a practice in the fashion accessories industry where designers' custom design jewellery and accessories are produced on a small scale, finished, packaged and distributed/ sold. Spectacle frames are already being manufactured by laser sintered method in titanium as seen in **Figure 14**. This may be extended to interior decors/decals and fittings that do not have to take weight and load.

Figure 16.
High-speed CNC machining of car body (source: cnc-modelle.com).

f. **Others:** With rapid physical model making moving well ahead of the initial curiosity, it could be left to the creativity of future designers, engineers and business people to derive further uses for rapid physical model making to suit the demands of the word.

6. Summary and conclusions

This chapter gives a historical background to physical model making in the profession of industrial design right from the last decades of the 19th century through the foundation and development of industrial design during the 20th century till the first two decades of the 21st century. Alongside the history, the different types of physical models, together with the reasons for using each type was explained before touching on the research and development of rapid physical model making technologies, specifically high-speed CNC, laser cutting and 3D printers. Examples of how they are used has also been given in this chapter before the author moves on to discuss on how rapid physical model making is taking the profession of industrial design to its next phase of progress.

What this chapter presents is an overview of the importance of physical model making and the role it has played before moving on to how rapid physical models are setting the scene for the future of industrial design, all within the context of the other chapters in this book.

Industrial design has played a major role in the modern and postmodern era, by bringing a harmonious relationship between form, function and aesthetics in the built environment and life style over the last 100 years. With the Earth facing imminent danger to its natural sustainability, it is hoped that the profession will take advantage of the technological advances brought about by CAD, rapid physical model making and its development in sustainable materials together with high speed connectivity to influence the direction of human habitat on this planet in the near future, and perhaps other celestial bodies in the future to come.

Author details

Peer M. Sathikh
School of Art, Design and Media, Nanyang Technological University, Singapore

*Address all correspondence to: peersathikh@ntu.edu.sg

References

[1] Bayley S, editor. The Conran Directory of Design. New York, USA: Villard Books; 1985. ISBN0-394-54698-9

[2] Loewy R. Industrial Design. Faber and Faber; London, UK: 1979. ISBN0-571-11470-9

[3] Broek JJ, Sleijffers W, Horvath I, Lennings AF. Using Physical Models in Design. CAID. The Netherlands: Delft University, Delft; 2000

[4] Isa SS, Liem A. Classifying physical models and prototypes in the design process: A study on the economical and usability impact of adopting models and prototypes in the design process. In: Proceedings of the International Design Conference (DESIGN 2014), Dubrovnik. 2014. pp. 2071-2081

Chapter 9

Effects of Dispersed Sulfides in Bronze During Sintering

Tomohiro Sato

Abstract

Bronze material sintered as sliding bearing is used. In particular, lead bronze is often used because lead acting as a solid lubricant has excellent friction characteristics. However, lead was replaced by another material according to environmental regulations. One candidate for a lead-free material is a sulfide that is well known as a solid lubricant. In this chapter we describe sintering properties and their mechanical properties. First, we investigate chemical components of copper sulfide system and realize stable phase in bronze matrix. After that, we consider the sintering condition of bronze with sulfide dispersed. The sulfides in the bronze may be subject to chemical reduction during sintering, especially when this is carried out under a reducing atmosphere containing hydrogen gas. The effect of the sulfides on the bronze, with a focus on the hardness of the bronze matrix and the reaction between sulfides and hydrogen gas, was investigated. Not only sinterability but also mechanical properties as hardness are discussed.

Keywords: sintering, Cu alloy, bronze, sulfide, hardness, friction

1. Introduction

Copper alloys based on copper and tin are useful materials as the so-called bronze (Cu-Sn alloy). For example, sliding bearings are common in industrial applications. These bearings are often manufactured using a sintering process in the field of powder metallurgy. Specifically, 90 Cu-10 Sn mass% bronze is common and has been developed and manufactured as a base materials of a bearing component.

Lead is a common element for use as industrial additives. This is also an indispensable material for additives for metal casting of Cu alloys and solders. The addition of lead ensures that the pressure resistance of the bronze and brass water devices and the mechanical properties of the solder are maintained. Lead is also useful in the manufacturing process. For example, lead made castability improve in the metal casting method [1, 2], and solderability is also improved due to low melting point of lead [3]. Lead has important applications as a solid lubricant, and lead bronze is a useful material for sliding bearings even though there are many regulations as RoHS and REACH and so on.

However, lead is harmful for human health, especially children and pregnant women. For that reason, various regulations are implemented to protect people.

These regulations or trends will affect industrial manufacturers. As a result, many industries have developed new materials as substitutes for lead.

Bismuth and sulfides are well known as lead substitutes and are candidates for the development of solid lubricants. Bi-bronze castings were used as bimetallic bearings and showed good loading capacity [4]. Potential applications of these castings are pointed out in Ref. [4], and it has been shown that castability was improved using Bi. The machinability of bronze-containing sulfides (Cu_2S and ZnS) was investigated. It was concluded that the mechanical properties and machinability of sand casting are the same as the mechanical properties and machinability of Pb-bronze castings [5]. For industrial, it is important that the manufacturing cost is low and that there is a stable supply of raw materials. Bi is far more expensive than S and Cu. Since Bi is a rare metal, the supply of sulfide seems to be superior to that of Bi. As a result, sulfide based on S and Cu is a promising alternative for lead substitution material.

In this study, we will discuss how to influence bronze sulfide which is already atomized. Specifically, the strength of bronze matrix and the reaction between sulfide and hydrogen gas are drawing attention. We clarify composition and atmosphere effective for sintering bronze with sulfide dispersed. In the investigation, solid-state sintering and liquid-phase sintering are compared under reducing atmosphere and inert atmosphere. Hardness as one of the important mechanical properties was also investigated. Moreover, some other sintering conditions and friction properties are also discussed.

2. Chemical component of sulfide-dispersed bronze

As shown in **Figure 1**, a phase diagram was calculated based on the calculation of phase diagrams (CALPHAD) method [6] of the Cu-Sn-Fe-S system in order to confirm the optimum content of the sulfide in the cast material. In the Cu-Sn alloy, since the crystallization of the α' phase of Fe was suppressed, the Fe content was 1.3 mass% or less. The reason why the sulfide was dispersed in the Cu alloy is that crystallization of the sulfide (0.25 mass% or less) occurred after crystallization of the α-phase Cu. On the one hand, the sulfide in the matrix remained below 0.60% by mass experimentally. **Figure 2** shows the matrix and sulfides of casting alloys [7].

On the other hand, much amount of sulfide is able to disperse in the atomized bronze powders. Because of the rapid cooling of (gas and/or water) atomization, atomized bronze keeps their metastable state including sulfide.

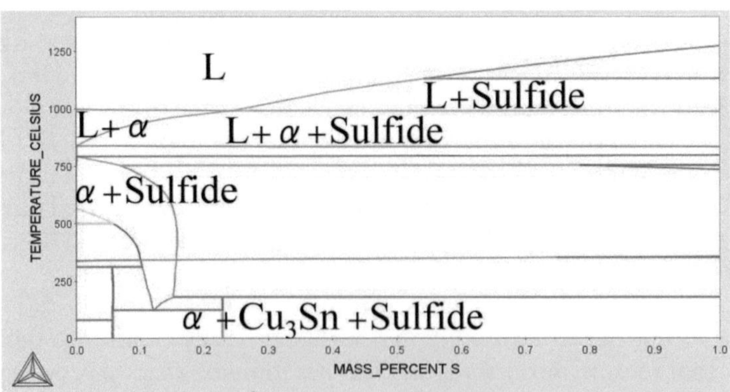

Figure 1.
Calculated phase diagram of sulfide bronze [7].

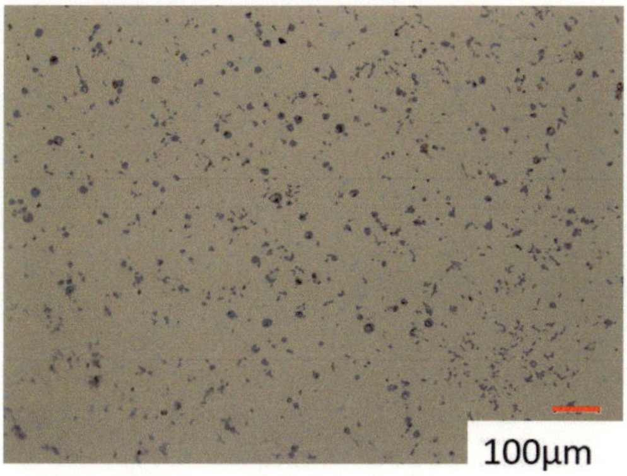

Figure 2.
Microstructure of sulfide bronze of castings.

3. Sintering process of sulfide bronze as bimetal

In this chapter, one of the conditions is the proposal of the copper alloy developed as the sliding member described below.

In the case of a Cu alloy containing a sulfide, sulfur may disappear from the Cu alloy by reacting with the reducing gas. Further, in the case of Cu-Sn-S, the mechanical properties of the sintered body have not been clarified. Therefore, hardness is evaluated as one of the mechanical properties.

3.1 Materials

Atomized powders were prepared for comparison of sintering properties. As a feature of the sulfide-dispersed Cu-Sn system materials, the sulfide was pre-alloyed by water atomization manufacturing. As shown in **Figure 3a–e**, micro-sized small dots were observed by scanning electron microscopy (SEM). This image is a sectional view of one of the typically sintered bronze-containing sulfides (from irregular powders). Generally, sintered composite from premixed bronze and sulfides indicated lower mechanical properties. It was reported that mechanical properties become better to cover the MoS_2 particles by copper [8]. However, this sulfide-dispersed bronze was made by atomizing as pre-alloyed material. So we can see that the pre-alloyed material shows better mechanical properties than the premixed material.

Energy-dispersive X-ray spectroscopy (EDS) was performed to determine the elements that make up the observed small dots. As a result, as shown in **Figure 1**, a ternary sulfide consisting of Cu, Fe, and S was detected. This sulfide is a kind of bornite (Cu_5FeS_4) detected by the X-ray diffraction (XRD) method [9] as shown in **Figure 4**. Only a small peak was observed as bornite (dot references). It may be metastable in the system because it is difficult to crystalize in the phase diagram as shown in **Figure 1**.

3.2 Experimental method

3.2.1 Sintering and manufacturing process

By preparing, bimetal specimens, these procedures are conducted as below.

At first, powder was sprayed to a height of 1.0 mm (by leveling off) onto a 3.2-mm-thick steel plate (low-carbon steel). At this time, binding materials such

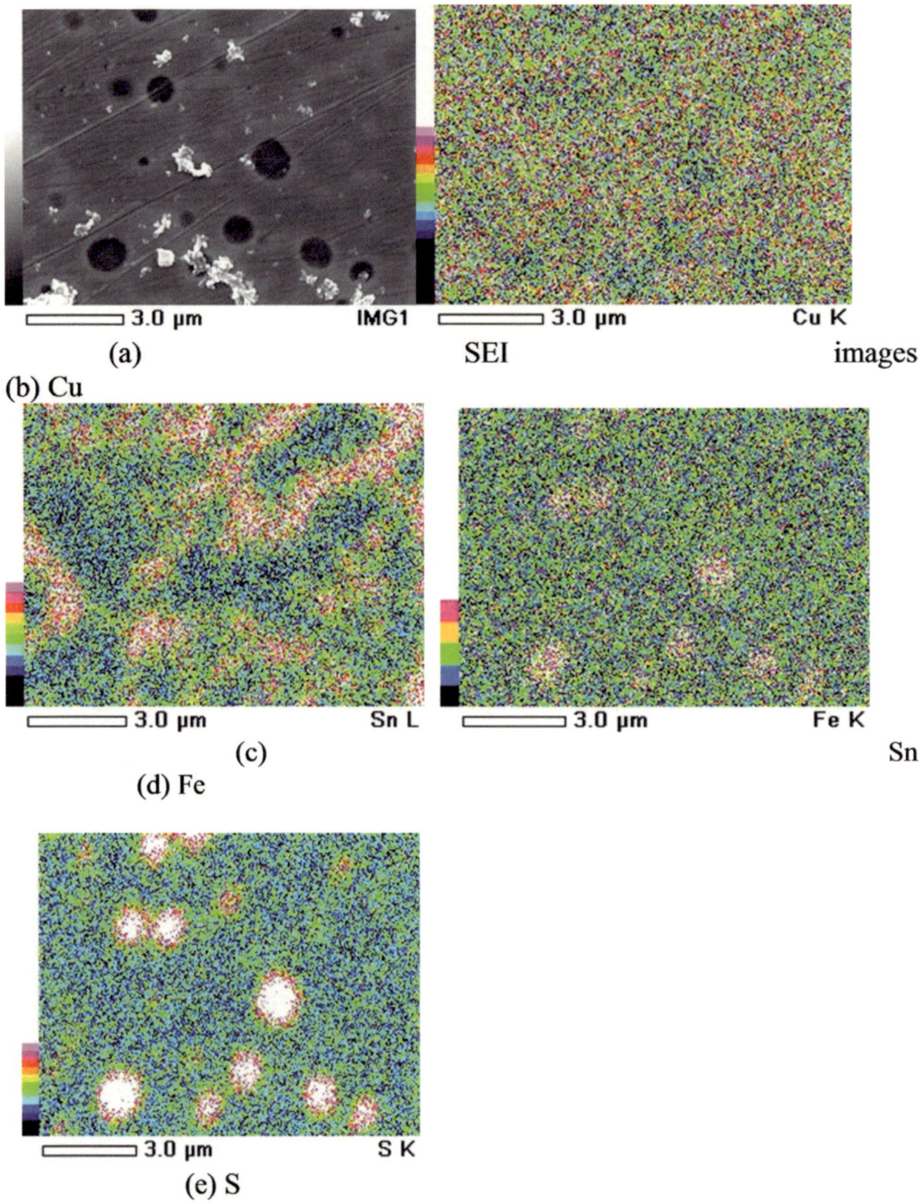

Figure 3.
SEM image of and EDS mapping results for the sulfide-dispersed copper alloy [10].

as oils or zinc stearate were not mixed. As the next step, the first sintering under reducing and/or inert atmosphere is performed.

The sample was heated to 1123 K for 1050 s in a mesh belt furnace. Next, the thickness of the bimetal was adjusted by a cold rolling mill. The thickness of the bimetal was controlled to flatten the entire surface on the Cu side of the bimetal in contact with the roll surface. Then, a second sintering under reducing and/or inert atmosphere is performed. The sample was heated to 1123 K for 1368 s in a mesh belt furnace.

After these procedures, to the analysis of mechanical properties, hardness tests were conducted because the bimetals and sintered copper alloy specimens were too thin to be subjected to tensile tests. The hardness of the sintered copper alloy specimens was evaluated using a Vickers hardness meter. This testing machine uses

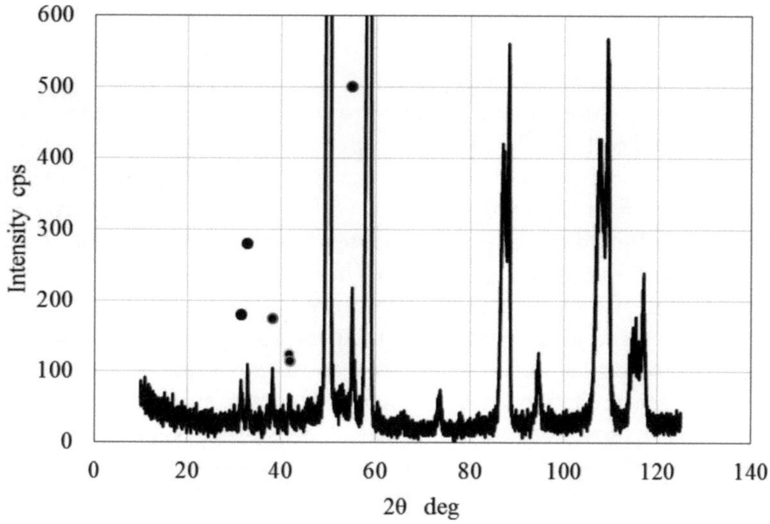

Figure 4.
XRD of sulfide-dispersed bronze powders.

a regular square pyramid diamond indenter with a face-to-face angle of 136° and a pyramid-shaped hollow on the test surface.

The hardness represented by a value obtained by dividing the load at that time by the surface area obtained from the length of the diagonal line of the permanent indentation is the Vickers hardness. The indenter was pushed into the matrix of the copper alloy and the steels under an applied load of 0.98 N for 10 s.

3.2.2 Test specimens

Six test specimens were prepared for observation of the sintering process, as shown in **Table 1**. In the table, the chemical components are described only for the primary element (i.e., Cu, Sn, Fe, S, and sulfide). Here, with respect to the sulfide concentration, all sulfur was assumed to exist as bornite.

As preliminary alloy powders for solid-phase sintering at the time of initial-stage sintering, SB8 (sulfide bronze containing 8 mass% Sn), SB10, and SB12 were used. However, SBP8 (sulfide bronze using a premix containing 8% by mass Sn), SBP10, and SBP12 are the same pre-alloyed powders as used in the first sintering (SB8, SB10, SB) (prepared as a mixture of 12). Bronze (containing 20% by mass Sn) powder. These mixed powders are prepared for liquid sintering at the time of primary sintering. The addition of low-melting Cu-Sn powder during the sintering process was tried experimentally to improve sintering [10].

Materials		Cu	Sn	Fe	S	Sulfide
SB	8	Bal.	7.90	0.18	0.31	1.21
	10	Bal.	9.52	0.38	0.48	1.88
	12	Bal.	12.00	0.41	0.58	2.27
SBP	8	Bal.	8.00	0.13	0.20	0.78
	10	Bal.	10.00	0.15	0.26	1.02
	12	Bal.	12.00	0.29	0.37	1.45

Table 1.
Chemical composition of powders primary elements (mass%) [10].

3.2.3 Test conditions

The sintering temperature was kept at 1123 K, as described in Section 3.2.1. This temperature made Cu-Sn system in liquid state so that liquid sintering was conducted for SBP8, SBP10, and SBP12. The mesh belt speed was set at 0.43 mm/s (first sintering) and 0.33 mm/s (second sintering). So, test specimens were maintained at 1123 K for 1050 s during the first sintering and for 1368 s during the second sintering. These sinterings were conducted under only reducing atmosphere (indicated as a after the materials in **Table 1**), only inert atmosphere (indicated as b after the materials in **Table 1**), and mixed atmosphere. Some tests were conducted using a complex procedure. They were sintered under inert gas during the first sintering and under reducing gas during the second sintering (indicated as c after the materials in **Table 1**). Under the reducing atmosphere, not only oxygen but also sulfur was able to reduce. Here, the reducing atmosphere was a mixture of H_2 gas and N_2 gas, and the inert atmosphere consisted only of N_2 gas. Sulfur in bronze may react as H_2S in the reducing gas. Thus, sintering under an inert atmosphere was also performed to compare the states of sulfur and sulfide bronze. Between the first and the second sinterings, a rolling process was performed, and the thickness of the bimetal was controlled to level the all surface of the Cu side of the bimetal contacted with the roll surface.

3.3 Results and discussion

3.3.1 First sintering

Figure 5 shows the results of optical microscopy observations. Specimens SB8a **Figure 5a**, SBP8a **Figure 5d**, SBP10a **Figure 5e**, and SBP12a **Figure 5f** were well sintered and contained few pores. Specimens SB10a **Figure 5b** and SB12a **Figure 5c** were difficult to observe because the sintered copper layer and steel were not adhered. Previous research has shown that the grain boundaries in copper alloys almost match the boundaries of individual particles (atomized powders); thus, diffusion may occur only at the surface of the particles because many sulfide dots remain in the Cu matrix. However, specimen SB8a had no grain boundaries. This is because of the

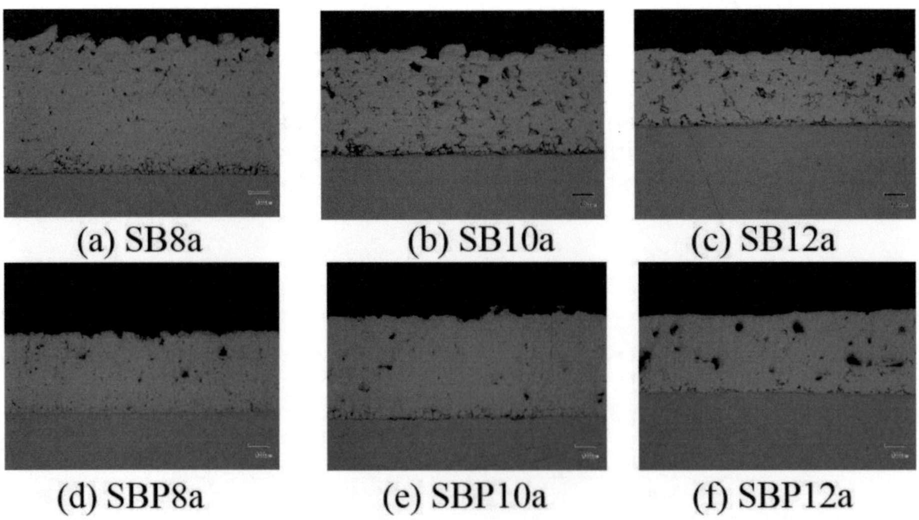

Figure 5.
Sectional view of sintered bimetal specimens after their first sintering under a reducing atmosphere [10].

differences between the sulfur content used in the previous study (S = 2 mass%) and that used in this study (S = 0.3 mass%).

In the cases of SBP8, SBP10, and SBP12, our results are similar to those reported in a previous study conducted under batch furnace conditions. Since the Cu-20 Sn powder begins to melt at 1071 K, effective diffusion is caused by liquid-phase sintering in Cu-Sn, since it is lower than the sintering temperature of 1113 K used in this study. **Figure 6a–e** shows the results of optical microscope observation of bimetal SB series and SBP series sintered in inert atmosphere. Samples SB8inr, SB10inr, SB10inr, SBP8inr, SBP10inr, and SBP12inr were well sintered and showed little pores.

In the sintered bronze, a dark brown network that appeared to comprise sulfide was present to a greater extent than in the specimens treated under a reducing atmosphere. At the reduction atmosphere, sulfur-containing sulfides in the copper alloy were apparently reduced and converted into H_2S gas in the furnace.

3.3.2 Second sintering under reduction atmosphere

Figure 7a–e shows the results of the bimetal after the second sintering under reducing atmosphere. Pores in the sintered copper alloy almost disappeared. However, as shown in **Figure 7b**, the alloy and the steel test piece did not adhere, and the sulfide almost disappeared from the test piece of the copper alloy.

In the second sintering process, the copper alloy samples were too sintered, which means that their pores almost disappeared. **Figure 8a–f** is a cross-sectional view of the test piece after the second sintering in an inert atmosphere. All copper alloy specimens were well sintered and almost void-free due to the rolling procedure performed between the first and second sinterings. The sulfide remained in the sample since the reaction in the sulfur in the sulfide is prevented because there is no reducing atmosphere.

3.3.3 Second sintering under complex atmosphere

Figure 9a–f is a cross-sectional view of a sintered bimetallic test piece after a second sintering in a reducing atmosphere and a sintered bimetallic test piece after

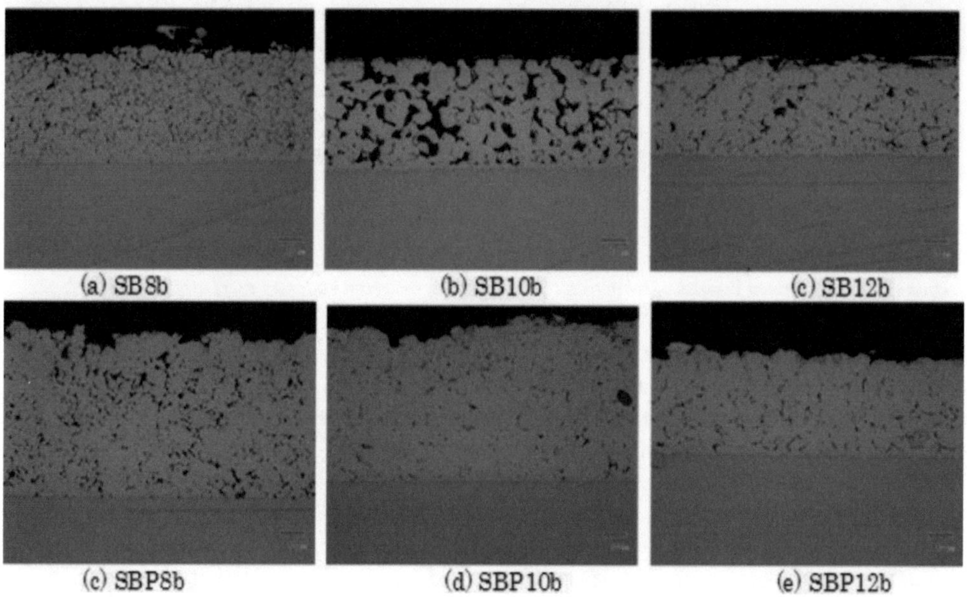

Figure 6.
Sectional view of sintered bimetal specimens after their first sintering under an inert atmosphere [10].

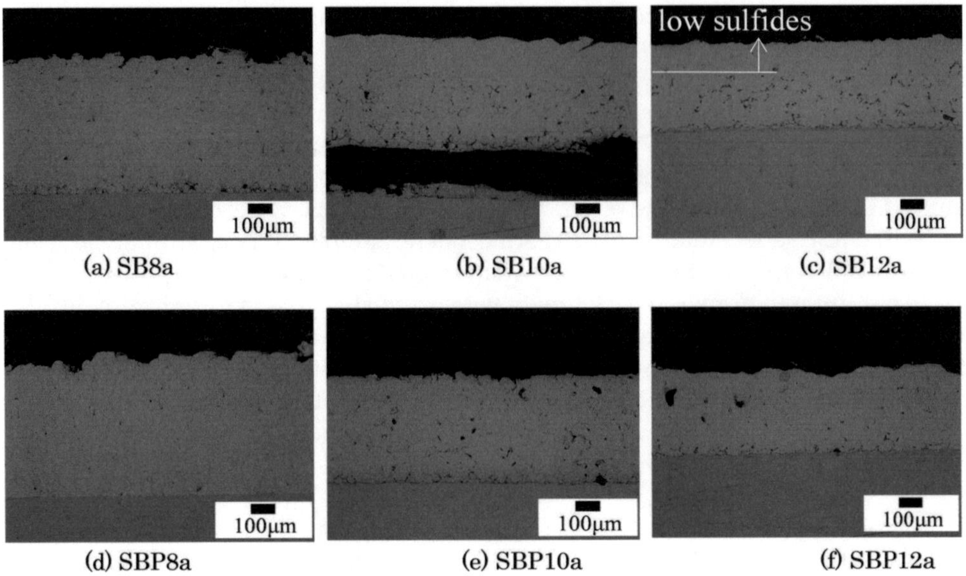

Figure 7.
Sectional view of sintered bimetal specimens after their second sintering under a reducing atmosphere [10].

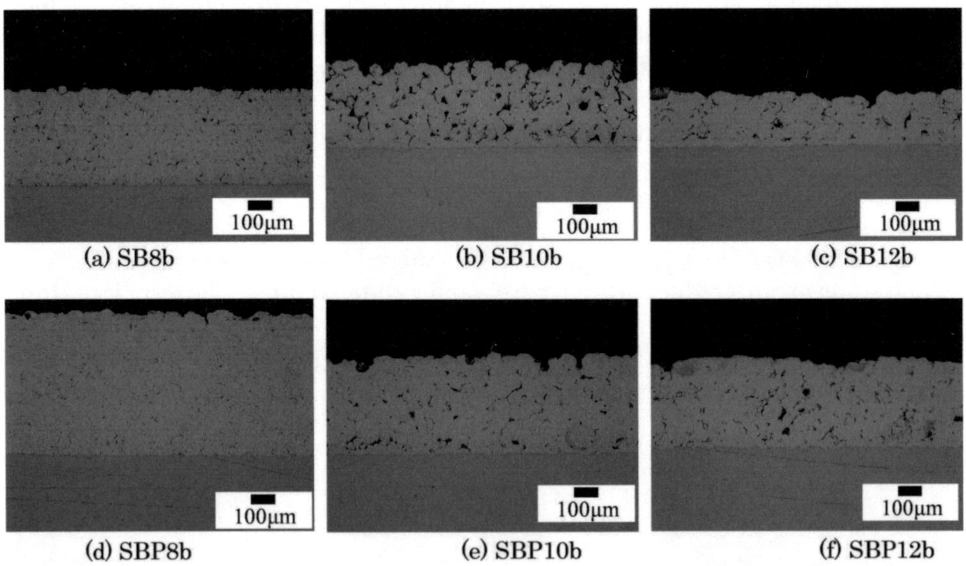

Figure 8.
Sectional view of sintered bimetal specimens after their second sintering under an inert atmosphere [10].

a first sintering in an inert atmosphere. The copper alloys in all the specimens were completely sintered and contained fewer pores than the specimens sintered only under inert atmosphere for both their primary and secondary sinterings. However, the sulfide content of the sintered bronze became inclined from the surface side to the interface side. So, a simple quantitative method of EDS was used for SPB10c.

As shown in **Figure 10a** and **b**, the sulfur on the front side disappeared, and in the simple quantitative method, only 0.2 mass% sulfur was detected. On the other hand, as shown in **Figure 10c** and **d**, the sulfur on the interface side remained, and 2.2% by mass of sulfur was detected. Therefore, the second sintering under reducing atmosphere resulted in the removal of sulfur from the surface of the copper alloy. Because the sulfide peak is small as shown in **Figure 4**, it

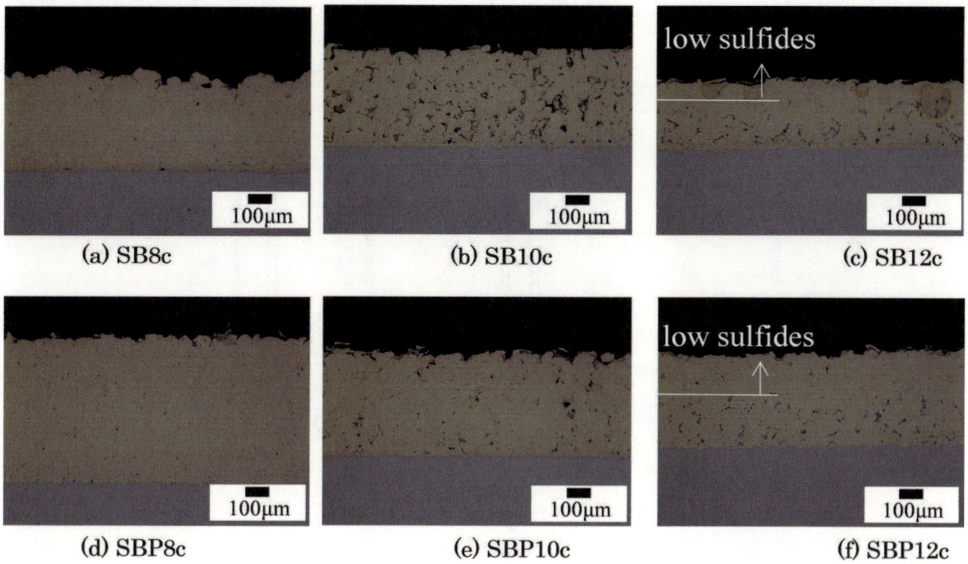

Figure 9.
Sectional view of sintered bimetal specimens after their second sintering under a reducing atmosphere [10].

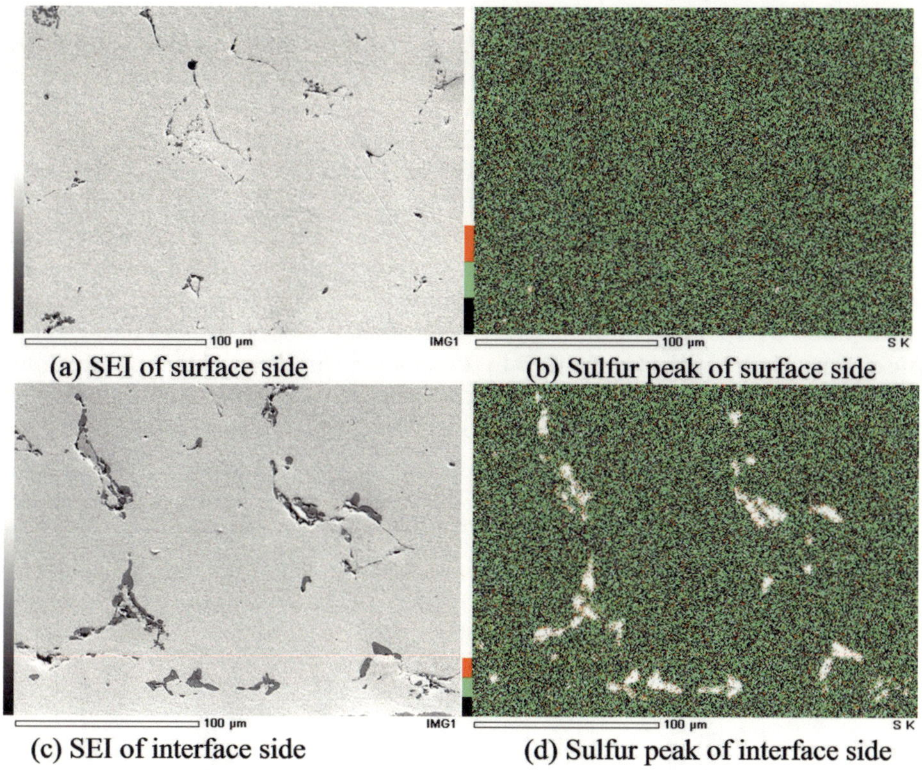

Figure 10.
SEI images and sulfur peak of SPB10c after their second sintering [10].

was difficult to detect sulfide by XRD in the cross section of bimetal. Therefore, apply EDS observation. For reference, XRD results (not cross-sectional views) of sintered bronze are shown in the figure. Bornite was also detected as shown in **Figure 11**. The figure shows the results of XRD for the sintered sulfide-dispersed bronze.

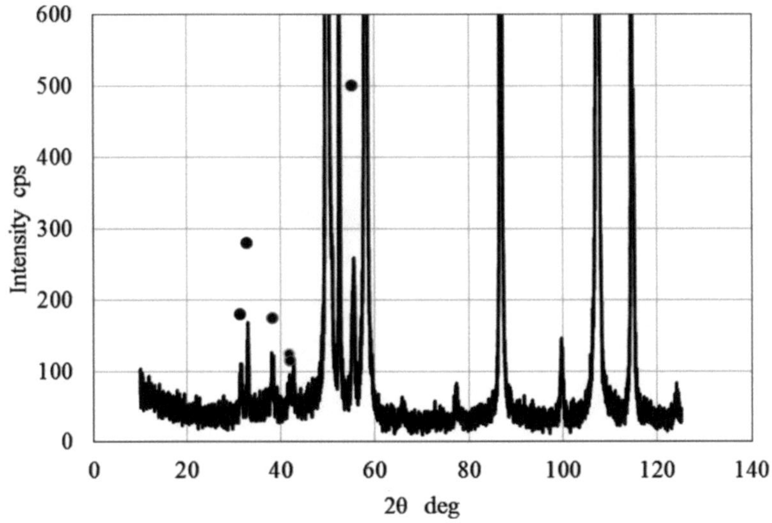

Figure 11.
XRD of sintered sulfide-dispersed bronze.

3.3.4 Hardness

Figures 12 and **13** show the Vickers hardness values of steel specimens and sintered copper alloys. As shown in **Figure 12**, the hardness of steel was almost the same during sintering and rolling. For steel, these thermal and operating conditions did not affect to change their microstructure and strength. It was important that the bimetallic bushing does not change the properties of the steel. As a result, manufacturing conditions were suitable for bimetal. As shown in **Figure 13a–c**, the results were divided into three categories depending on the sintering atmosphere. For all of the test groups that received their initial sintering, the high Sn content alloys are harder in each group. It is known that the Cu-Sn alloy becomes hard when the Sn content increases within the range of the Sn content of the test piece examined in this study. As a result, the basic characteristics of the Cu-Sn alloy in the sintered Cu alloy and the sulfide content in the Cu alloy had almost no influence on the hardness of the base material. After the rolling step, some samples became harder, but the other samples were not hard.

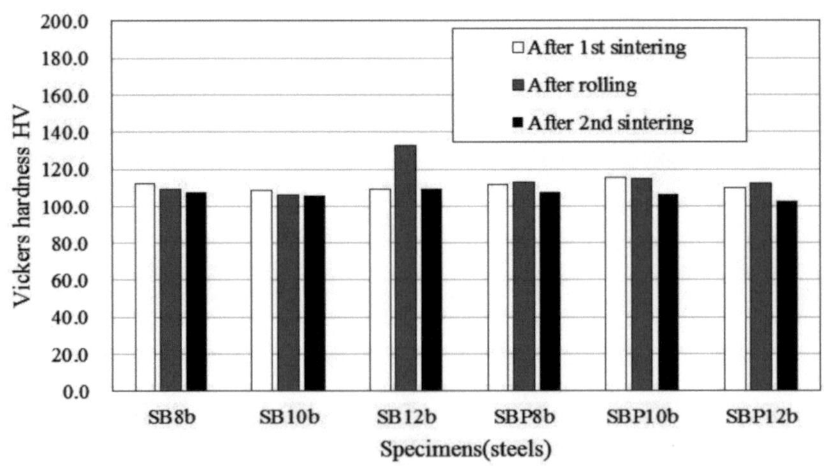

Figure 12.
Vickers hardness of steels under inert atmosphere [10].

The increase in hardness of some specimens may be the result of work hardening during rolling, and the lack of increase in hardness of other specimens may be a result of the destruction of the interface formed by solid diffusion between the powders.

When the hardness of SB group and SBP group after the second sintering is compared, Sn group is harder than Sn content (viz., 8, 10, 12 mass% Sn). After the second sintering, Cu_3Sn which was one of the intermetallic compounds was precipitated or crystallized. In this series, liquid sintering may occur during sintering. As a result, matrices with low tin content become softer after sintering.

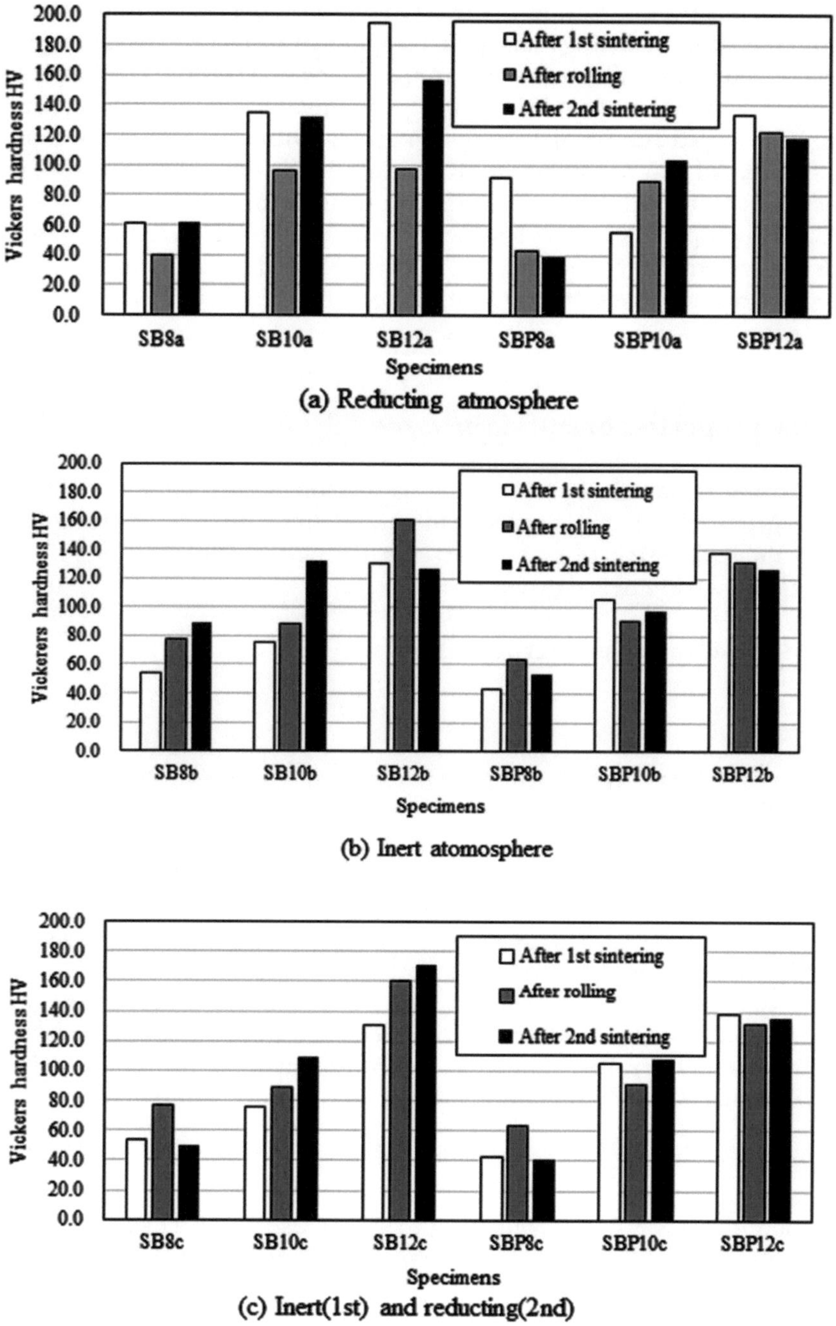

Figure 13.
Vickers hardness of Cu matrix [10].

Moreover, some specimens only sintering under inert atmosphere shown in **Figure 13b** were not harder than specimens only sintering under reduced atmosphere shown in **Figure 13a**, relatively. Reducing atmosphere might progress sintering under the same sintering temperature.

From these results, by using pre-alloyed atomizing bronze-containing sulfides, decreasing mechanical properties as hardness were not observed.

3.4 Summary

The effect of sulfur or sulfide on the mechanical properties of bronze specimens sintered under reducing inert gas atmosphere was investigated by subjecting water-atomized sulfide-dispersed bronze specimens to sintering. Furthermore, solid-phase sintering and liquid sintering were compared for the same Sn content. By using pre-alloyed atomizing bronze-containing sulfides, no mechanical properties were observed that decreased significantly as hardness. The sulfide content of the specimens decreased during sintering in a reducing atmosphere. With regard to mechanical properties such as hardness, the Sn content affected the properties regardless of whether the specimen had undergone solid-phase sintering or liquid sintering. In contrast, the sulfur and sulfide content and mechanical properties had no correlation.

4. Friction properties of sulfide bronze

In this study, tribological properties of copper alloys with sulfide particles are discussed. Friction characteristics were measured under dry and lubricated conditions to evaluate the effect of sulfide particles. Graphite penetrating into the pores caused improved tribological properties [11].

4.1 Materials and experimental procedure

4.1.1 Materials

The developed lead-free material is based on bronze-type Cu-Sn alloy and contains sulfide. For friction materials that use a sintering process, sulfides are considered to be an alternative element to lead, tested as TP-L (Cu-10 Sn-10 Pb mass%: lead bronze) and TP-A1 (Cu-12.15 Sn-1.78 Fe-0.48 S mass%) and for the purpose of comparing tribological properties using new materials prepared as TP-B1 (Cu-11.93 Sn-1.44 Fe-1.78 S mass%). The test specimen is sintered and rolled onto a steel plate, and the powder is separated. This plate was heated in furnace under reduction condition. In order to sinter to make the bimetal, the powder on the plate has to be heated at 1113 K for 10 min. After sintering, reduce the thickness of the plate by rolling, then sinter the rolled bimetallic, and roll again under the same conditions used in the initial.

4.1.2 Friction test

The friction test was performed using a ring on disk tester. The contact load was applied by its own weight and was in the range of 50–600 N. Disk specimens were attached to a shaft driven by a DC motor. The sliding speed could be controlled continuously in the range of 0.1–1.4 m/s. Mineral oil (approximately 30 μL, three drops from a microsyringe) is delivered to the interface just prior to testing. The disk surface was first contacted with the ring, and then the disk specimen was driven.

The laboratory air and the laboratory environment under the break-in process had a load of 20 N and were applied at a sliding distance of 120 m before testing. These specimens and carbon steel disks (S45C) were mirror finished. The roughness was Ra 0.025 μm (TP-L), Ra 0.167 μm (TP-A), Ra 0.343 μm (TP-B), and Ra 0.004 μm (carbon steel).

After interrupting at 20 N for 10 min in operating mode (friction distance 120 m), gradually load the ring in 50 N increments until a friction coefficient of 0.2 is reached or strange noise occurs. In the lubricant test, PAO (50 cSt @ 313 K) was used as a lubricant. Another test was performed on this material for high speed and low load [12].

4.2 Results and discussion

Friction coefficient during the test was shown in **Figure 14**. In the beginning of the test, lower friction coefficient of around 0.05 was measured in TP-B.

Friction coefficient of TP-L also shows a low coefficient of friction at the beginning of the test. After that, the friction coefficient abruptly increases at a load of 300 N. The friction coefficient between TP-A and TP-B is reduced at 200 and 300 N load. From the result, these sintered specimens could achieve a lower coefficient of friction.

Lubricating oil seems to hinder the formation of S-based coatings. Due to the pores passing through the lubricating oil, the difference in coefficient of friction is not so wide between the dry test and the lubricating oil test.

From **Figure 15a–d**, surface state after the test was observed. Attachment to the ring (carbon steel/S45C) of the copper alloy particles (TP-B) which is a part of the disk is observed (see the circle in **Figure 15b**). On the other hand, adhesion of the part of carbon steel to the disk was observed as shown in **Figure 15c**. The adhesion of the Cu alloy to carbon steel achieves a lower coefficient of friction than the adhesion of carbon steel to Cu alloy at high loads. These phenomena also reported another kind of friction test as sulfides play a role as solid lubricants [13].

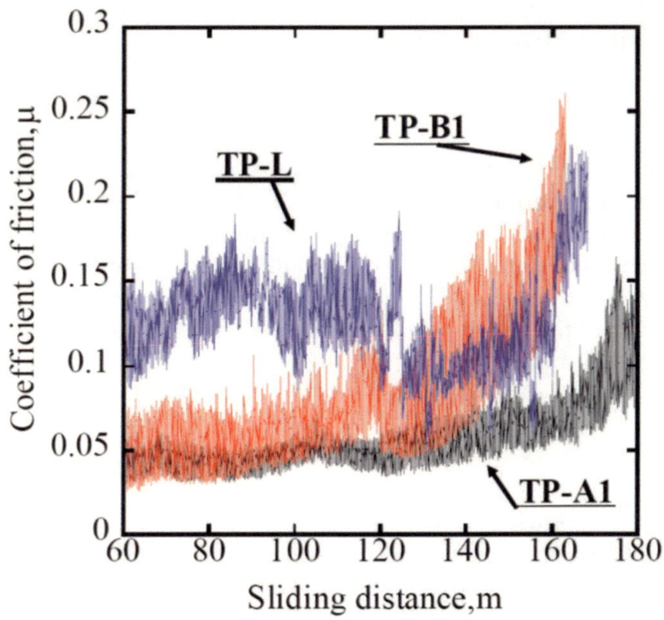

Figure 14.
Results of friction test [11].

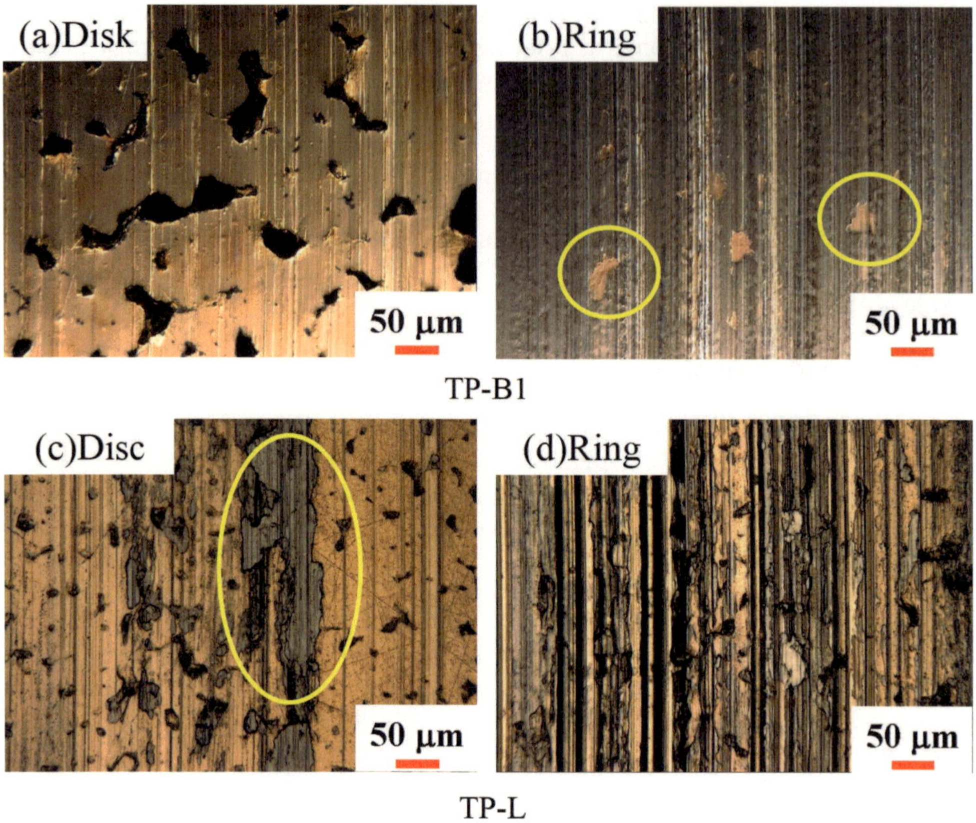

Figure 15.
Surface of disks and rings after the test [11].

In particular, MoS_2 was a common solid lubricant. Mixed MoS_2 in the Cu-based sintered material had better friction characteristics [14]. The sliding mechanism of MoS_2 is well known, and it is possible that a born person has the same or similar sliding mechanism as MoS_2 [15–17].

4.3 Summary

A friction test was conducted on the developed sintered Cu alloy-containing sulfide particles. The friction performance of the developed material was compared with those containing lead. The following is a summary of the results obtained.

The change in coefficient of friction depends on the behavior of the adhesive layer, especially under high load.

Sulfide-dispersed porous sintered Cu alloy realizes stabilization of friction coefficient at higher load than lead bronze.

5. Conclusions

In this study, we discussed how it affects already atomized bronze sulfide. Specifically, the strength of bronze matrix and the reaction between sulfide and hydrogen gas are drawing attention. We will clarify the composition and atmosphere effective for sintering bronze with sulfide dispersed. As a result, copper sulfide can sinter several types of atmosphere such as inert, reducing, and vacuum. In many reducing conditions, attention is paid to the disappearance of sulfides from the matrix. However, the sulfide itself did not affect the hardness of the matrix. Sulfide works as a solid lubricant instead of lead because it shows high seizure resistance during friction testing.

Conflict of interest

The authors declare that they have no conflicts of interest.

Author details

Tomohiro Sato
Kansai University, Suita-shi, Osaka, Japan

*Address all correspondence to: tom_sato@kansai-u.ac.jp

References

[1] Maruyama T, Abe H, Hirose K, Matsubayashi R, Kobayashi T. Influence of alloying elements on sulfide formation in lead free bronze castings with dispersed sulfide particles. Materials Transactions. 2012;**53**(2):380-384

[2] Taha MA, E-Mahallawy NA, Mousa TM, Hamouda RM, Yousef AFAG. Microstructure and castability of lead-free silicon brass alloys. Materialwissenschaft und Werkstofftechnik. 2012;**43**(8):699-704

[3] Dudek MA, Sidhu RS, Chawla N. Novel rare-earth-containing lead-free solders with enhanced ductility. The Journal of the Minerals, Metals & Materials Society. 2006;**58**:57-59

[4] Oksane VT, Lehtovaara AJ, Kallio MH. Load capacity of bismuth bronze bimetal bearing in lubricated conditions. Proc. of the 17th Nordic Symposium on Tribology. 2016;**222**:1-13

[5] Maruyama T, Wakai H, Kobayashi T. Some properties of sulfide dispersed lead free copper alloy castings, AFS transactions. Transactions of the American Foundry Society. 2008;**116**:299-307

[6] Lukas H, Fries SG, Sundman B. Computational Thermodynamics: The Calphad Method. New York, NY, USA: Cambridge University Press; 2007

[7] Sato T, Hirai Y, Kobayashi T. Development of lead free bronze with sulfide dispersion for sliding members. International Journal of Metal Casting. 2017;**11**(1):148-154

[8] Shikata H, Funabashi N, Ebine Y, Hayasaka T. Performance of sintered Cu-Sn-Ni bearings containing MoS_2. Modern Developments in Powder Metallurgy. 1985;**17**:497-512

[9] Sato T, Hirai Y, Maruyama T, Kobayasi T, Kurimoto, Ltd. Copper alloy for sliding materials. US Patent 8906129; 2014

[10] Sato T, Hirai Y. Effects of sulfur and tin contents on hardness of copper-tin alloy under reduced atmosphere in sintering process. Mechanical Engineering Journal. 2016;**3**(1):15-00704

[11] Sato T, Hirai Y, Fukui T, Tanizawa K, Usami H. Tribological properties of porous Cu based alloy containing nano sized sulfide particles. Journal of Advanced Mechanical Design Systems and Manufacturing. 2012;**6**(1):158-167

[12] Sato T, Hirai Y, Fukui T, Akiyama K, Usami U. Effects of dispersed sulfides in bronze under line contact conditions. Jurnal Tribologi. 2016;**8**:1-11

[13] Sato T, Hirai Y, Maruyama T, Kobayashi T. Sintering processes of Pb-Free copper alloy for friction materials. In: Proceedings of Powder Metallurgy World Congress and Exhibition 2010; 10-14 October 2010; Florence. Vol. 2. EPMA; 2010. pp. 313-319

[14] Kovalchenko AM, Fushchich OI, Danyluk S. The tribological properties and mechanism of wear of Cu-based sintered powder materials containing molybdenum disulfide and molybdenum diselenite under unlubricated sliding against copper. Wear. 2012;**290**:106

[15] Hirai Y, Sato T, Usami H. Combined effects of graphite and sulfide on the tribological properties of bronze under dry conditions. Journal Tribologi. 2016;**11**:14-23

[16] Hirai Y, Sato T, Usami H. Tribological properties of sintered

bronze containing micro-sized Sulfide. Journal of Japanese society of Tribologists. 2016;**61**:857-865

[17] Hirai Y, Ogawa K, Sato T, Usami H. Effect of machining history on tribological properties of bronze containing micro-sized Sulfide. Key Engineering Materials. 2017;**749**:246-250

Section 3

Biomedical Applications

Chapter 10

Comprehensive Review on Full Bone Regeneration through 3D Printing Approaches

Cristiana Fernandes, Carla Moura, Rita M.T. Ascenso, Sandra Amado, Nuno Alves and Paula Pascoal-Faria

Abstract

Over the last decades, the number of work accidents associated with bone fractures has increased leading to a growing concern worldwide. Currently, autografts, allografts, and xenografts are used for bone regeneration. However, their application has associated risks. Tissue engineering (TE) has brought solutions to address these problems, through the production of temporary supports, providing mechanical support to the formation of new bone tissue and biocompatible and biodegradable scaffolds, which allow cell adhesion and proliferation to ensure bone formation. The combination of materials and structure with the technique to be used will directly influence their physical and chemical properties and, consequently, their action in contributing to bone regeneration. Thus, the focus of this chapter is to perform an exhaustive literature review and a critical analysis of the state of the art in bone TE and present a proposal of an optimized temporary support geometry for bone regeneration in case of large bone defects. For this, it was listed and identified the best choice of biomaterials, fabrication method, cell type and their culture conditions (static vs. dynamic), and/or the inclusion of growth factors for the repair of large bone defects.

Keywords: large bone defects, bone regeneration, tissue engineering, cell culture, CAD, scaffolds, additive manufacturing

1. Introduction

The population is increasingly exposed to accidents, both in daily routine and at work. In Portugal, among 209,390 non-death accidents that occurred in 2017, almost 4% were bone fractures that are limitative for the active population and require a long time of recovery [1]. Many research groups have been working on bone regeneration for over 10 years, but this has not led to effective therapy in a clinical setting. If it was successful, it would enhance the quality of life for millions of people and significantly reduce the absence to work due to fractures which are considered the second higher cause of working day lost.

The bone is a natural composite containing organic components (mainly collagen type I and fibrillin) and inorganic crystalline minerals (such as hydroxyapatite (Hap)), defined as hard tissue [2–4]. The characteristic of the collagen fibers in their structure gives it high tensile strength and its mineral substances impart high compressive strength and thus excellent mechanical resistance.

Bone, namely in the diaphysis, is made up of cortical or compact bone that contains its own blood vessels and cells, which aid in its growth and regeneration. It has many types of cells, such as osteoblasts, osteocytes, osteoclasts, and a matrix of non-mineralized collagen (osteoid). Bone tissue comprises several functions, such as: (i) provide structural integrity, and all the necessary support to the soft tissue of the body, constituting the global support of the majority of the muscles, (ii) protect vital organs, and (iii) help to balance the minerals, since the bone tissue stores calcium and phosphate making them more resistant and able to maintain a balance of blood concentration [5–7]. Bone is known to self-regenerate: pos-natal bone maintains an intrinsic capacity for well-ordered growth, remodeling to meet mechanical needs, and renewal after damage [8].

Major bone defects are the result of injury, trauma, nonunion after a fracture, infection, or abnormality, resulting in long-term deformities, such as limb shortening, leaving patients with reduced bone structure and function [9–11]. It should be noted that the most transplanted tissue after blood is bone [12, 13].

The gold standard treatments for bone defects are still bone grafts. These can be used alone or combined with other materials in order to promote bone healing through osteoinduction, osteoconduction, and osteogenesis [14]. These bone grafts may be from autograft (taken from the patient), allograft (taken from another patient) and xenograft (obtained from an animal) origins or even manipulated with synthetic biomaterials. Additionally, prostheses can also be used, but they usually need a second surgery later on due to some complications that may appear, such as the formation of bone callus or hernias. Autografts are still considered the gold standard treatment due to their osteogenic, osteoinductive, and osteoconductive capacity. However, there is a limitation in tissue extraction from the amount that is required. Allografts taken from other donors or corpses present a high risk of immune rejection, reduced bioactivity and a high risk of pathogen transmission [7, 15–18].

When there are fractures with a bone defect exceeding a critical size, the bone is not able to self-regenerate and, therefore, requires the use of a temporary implant (natural and/or synthetic) to serve as support and cells to help bone regeneration [19]. In this way, tissue engineering (TE) has emerged [20].

The concept of TE was implemented in 1993 by Langer and Vacanti. They specified that "TE is an interdisciplinary field that relates the principles of biology and engineering to the production of tissue functional substitutes" [21]. So, they presented specific characteristics and applications in biodegradable three-dimensional (3D) scaffolds. Ideally, they should be highly porous, having highly interconnected pore networks with a pore size suitable for cells to migrate and differentiate whenever necessary [22]. However, the biggest challenge of scaffolds is related to mass transport of nutrients and secretion of waste in tissue [6]. It is important cells used in 3D cultures of scaffolds be able to mimic the morphology, functionality, and biology of the tissue. These cell cultures are necessary to analyze mechanisms of chronic diseases and the impact of drug treatments or to produce different tissues for major defects in vivo, in this study, the bone. Bioreactors appeared to improve the field of cell culture on 3D support [23, 24].

This chapter intends to perform a critical analysis of the state of the art regarding full bone TE towards the selection of the most appropriate solution of temporary implants. Thus, the optimum conditions (static vs. dynamic), material, cells, and/or the inclusion of growth factors for the repair of large bone defects are discussed. Hence, there are two scientific questions to which this chapter intends to address: (i) which is the most suitable combination of scaffold design and fabrication using a certain biomaterial and biological components to facilitate or accelerate bone regeneration and (ii) what are the in vitro conditions more suitable to achieve an optimized in vivo response.

2. The usefulness of temporary implants for bone regeneration

Bone tissue is known for its ability to self-regenerate on its own. However, if the fracture becomes a critical bone defect, the bone loses this ability. From 1934 to the present day, some authors argue that a bone defect becomes critical when it is over two times the diameter of the bone defect [10, 11, 25–27].

These critical-sized defects may result from infection, malformation, and traumatic injuries, which may lead to bone loss in the patient [28–33]. In this case, as bone cannot self-regenerate, it is necessary to use a temporary implant (natural and/or synthetic) to support bone regeneration with cellular incorporation. To achieve this, successfully, it is first necessary to consider the mechanical properties of the native bone tissue.

It is known that the mechanical properties of the bone vary according to age, anatomical location, and bone quality. Within the biomechanical properties of the bone (resistance, stiffness, and fatigue), the elastic modulus is the most attracting variable in research due to its importance to characterize bone pathologies and also in the design orientation of artificial implants. Bone strength and elasticity are anisotropic. The compact bone is stronger under compression and stiffer when loaded longitudinally along the diaphyseal axis than in the transverse radial directions. In trabecular bone, its mechanical properties depend on both the porosity and the architectural desirability of the individual trabeculae [3]. The mechanical properties of human bone are summarized in **Table 1**.

Implants need to be accepted by the human body, where there are guarantees for cell survival in a safe and supportive environment. Moreover, mechanical damage or failure caused by stress shielding must be prevented. The scaffolds need to have an appropriate modulus of elasticity to match bone properties. Scaffolds with a highly porous structure are favorable for cellular activities, including fixation and proliferation, which will contribute to bone neoformation and regeneration and adjust the mechanical properties in terms of Young's strength and modulus [36].

So, an important key factor is concerned with the type of materials to be used in the implant. Biopolymers are biocompatible and biologically active materials as they promote cell adhesion and growth.

To help in the engineering of long bone fracture regeneration, artificial fractures are typically manufactured in models in vivo [23, 37]. Various animal models are studied in vivo before its application in humans.

Human bone	Trabecular	Cortical
Porosity (%)	50.00–90.00	1.00–20.00
Young's modulus E (GPa)	0.05–0.10	17.00–20.00
Compressive strength (MPa)	5.00–10.00	131.00–224.00
Tensile strength (MPa)	1.50–38.00	35.00–283.00
Elongation at break (%)	0.50–3.00	1.07–2.10
References	[2, 3, 34, 35]	

Table 1.
Human long bone properties.

3. In vivo studies

The bone, in vivo, is exposed to mechanical stimulation by muscle contraction and body movements, and the mechanical load induces an increase in bone mass formation [38]. During body movement, the forces applied results in changes in hydrostatic pressure, fluid flow-induced shear stress, direct cell strain, and electric fields [38–40].

In order to identify the mechanical properties necessary for humans, it is important to study what kind of in vivo studies and which animal models have been considered in the literature (see **Table 1**). The choice of the most appropriate animal model is an important step in clinical translation, because it will help to better understand and propose innovative strategies for bone regeneration. Each animal model has pros and cons [41], and in each study, a specific set of parameters is used. That is why it is difficult to compare the different studies available [42, 43]. There are various models that were studied in vivo for full bone regeneration. Rabbit, rat, ovine (sheep or bovine), canine, and goat are the most used.

There are six studies on the literature where a rabbit model was considered. Nather and their co-workers [44] evaluated the effect of bone marrow mesenchymal stem cells (BMMSCs) on the biological healing of a 1.5 cm cortical bone allograft in the tibia of adult rabbits. In their study it was shown that BMMSCs can improve cortical allograft binding rate, reabsorption activity, bone formation, and osteocyte cell count. In 2013, Khojasteh et al. [45] developed a scaffold using particulate mineralized bone/fibrin glue/mesenchymal stem cells (MSCs). Through the alizarin staining method, they verified that there was a deposition of mineralized matrix. This was also demonstrated by RT-PCR analysis of osteocyte markers. At the end of 3 weeks, osteocalcin, osteopontin, and collagen I messenger RNA were produced. They concluded that this implant would be a promising combination for vertical bone augmentation around implants inserted simultaneously into the tibia of rabbits. Lee et al. [46] studied the effect of autologous BMMSCs seeded into gel foam on structural bone allograft healing in 1.5 cm femoral defect of white rabbits. They concluded that the use of MSCs influenced the bone formation, resorption, and angiogenesis. Jang et al. [47] extruded porous HAp scaffolds, which were set in a drill-cut femur rabbit bone. After 4 and 8 weeks of implantation, micro-CT scanning images showed material degradation and integration of the sample into the native bone. In this period, the morphological behavior was similar in bone tissue-scaffold junction. Chowdhary et al. [48] had evaluated the early response of bone tissue to micro threads with an oxidized titanium implant (4 mm in diameter and 8 mm in length) between the macro threads. The study was tested in rabbit legs, tibia, and femur. The bone regeneration happened near the micro threads, and the bone growth in femur indicated that the cancellous bone seems to be more sensitive to micro thread stimulation. Recently, in 2018, Tovar and co-workers [49] used 3D printing, specifically robocasting/direct writing, to develop a scaffold with 100% beta-tricalcium phosphate (β-TCP) (350 μm pore diameter) in order to regenerate critical-sized rabbit radius defects in vivo. A 3 cm incision was made in a critical defect of 11 mm, approximately, in the radio, and the periosteum was resected to at least 1.5 cm proximal and distal to the defect. This scaffold proved to be good for bone tissue engineering (BTE) since at 8 weeks it showed bone formation with signs of resorption of the scaffold. The amount of bone formed was increased from week to week, regenerating the medullary space, and at 24 weeks the scaffold was significantly resorbed.

Rats are also another in vivo model referred in the literature. Saravanan et al. [50] introduced in an albino-Wistar rat with a critical-sized bone defect in the tibia a scaffold containing chitosan, gelatin, and graphene oxide by freeze drying.

They concluded that this scaffold promoted increase in osteoblasts and increased the collagen content, accelerating the bridging of the rat tibial bone defect.

A model of an ovine with 35 mm tibial defect was then used by Smith et al. [51] to study full bone regeneration. They produced a scaffold using blending process of poly (L-lactic acid)-poly(ε-caprolactone) (PLLA: PCL 20:80) with and without marrow-derived skeletal stem cells. They divided the tests into three different groups: empty defect, scaffold alone, and scaffold with cells. Radiographic has shown poor bone formation upon 12 weeks. However, there was a greater tendency for bone formation in the scaffold with cells.

Regarding the canine model, two studies were published in the literature. In 1996, Bragdon and co-workers [52] showed that, in canine femurs, an oscillating motion of 20 μm does not affect osseointegration. However, 40 and 150 μm oscillating motion of implants does not support bone growth. Recently, Barba et al. [19] implanted in vivo, in a canine model calcium-deficient scaffold (0.3 mm height and 5 mm diameter) with Hap considering different pore architectures and compared with two ceramics, a biphasic calcium phosphate (BCP) and a β-TCP with rat mesenchymal stem cells (rMSCs). Pores ranged from 10 to 300 μm. With this, calcium-deficient scaffold and Hap triggered osteogenic differentiation of rMSCs. They concluded that calcium-deficient HAp foam scaffolds with a spherical concave macroporosity allow osteoinduction.

Animal studies are needed to understand bone regeneration. Variables such as the amount of bone formation and its kinetics, mechanical properties and safety obtained by the scaffold, including the presence of toxic degradation in different organs and in terms of inflammatory response need to be understood in detail [42]. However, bone fractures performed in animals do not represent the complexity of healing human fractures [23, 37]. The potential of each different type of cells both in vitro and in vivo plays here a key role.

Ko and co-workers studied the potential of human-induced pluripotent stem cells (hiPSCs) against the human bone marrow mesenchymal stem cells (hBMMSCs). Both cells were placed in rat bone defects, with a size of 2 cm, which is similar to the human value mentioned above. They concluded that both hIPSCs and hBMMSCs have osteogenic potential in vivo [53]. However, some authors showed the existence of risks of teratoma formation after transplantation in hIPSCs [28, 54–56]. In literature, the use of MSCs seems to significantly help bone regeneration in in vivo studies [31, 45, 57–59]. Some authors defend that the addition of growth factors to cell-scaffold constructs promotes bone regeneration [60]. Nevertheless, Kleinhans et al. [61] showed that a good culture capable of mimicking tissue morphology, functionality, and biology, for example, using bioreactors, is sufficient to obtain a homogeneous cell distribution of soluble factors.

There is a great deal of discussion today about the incorporation of growth factors. In this chapter, authors defend the nonnecessity of its incorporation, since upon the right environmental conditions, cells are actually able to secrete the optimal extracellular matrix (ECM) components. Therefore, a good mechanically stimulated culture combined with transcription factors influences cells to bone formation.

4. In vitro studies

In vitro models are required to accurately record the physiology of healing at a site of bone fracture since bone takes weeks to differentiate in vitro [62] and wound healing can take weeks to months [23, 63]. In vitro studies are advantageous because they offer a controlled environment to experimental test molecular and cellular hypotheses. However, cells cultured in vitro are not replicates of their in vivo counterparts [28, 64].

When the bone is subjected to a mechanical force, electrical potentials are generated, which play an important role in bone remodeling. To mimic this natural process, bioreactors were created and are nowadays widely used. These 3D systems allow the control of various parameters, such as temperature, pH, oxygen concentration, growth factors, and mechanical stimuli, among others, and modulate cell growth more easily. These bioreactors can simulate the human bone environment and allow the study of the role of various factors in scaffolds or preculture scaffolds in vivo. In addition, to provide adequate nutrition and removing residues from all cells in the scaffold, fluid flow can be manipulated to physically stimulate bone growth [39]. Bone is constantly exposed to mechanical stimulation due to muscle contractions and body movements that result in changes in hydrostatic pressure, direct cell strain, fluid flow-induced shear stress, and electric fields. In addition, bone cells are more sensitive to mechanical stimulation. Therefore, providing physical stimulation in bioreactors becomes a key component of BTE strategies [65].

The following studies demonstrate the importance of performing in vitro testing in order to find the best strategy.

Jang et al. [66] developed a HAp scaffold to mimic native bone through a multipass extraction process with the addition of osteoblast-like cells, with pores of 150 ± 20 μm in diameter and with a pore structure of 50 ± 10 μm which is thin enough for rapid bone resorption. With in vivo tests and in vitro tests, they confirmed that the scaffold used is appropriate for graft without inflammatory reactions and bone formation after 8 weeks of implantation. The scaffold's porosity is a critical parameter enabling medium exchange and nutrient diffusion, which is a key role in cell proliferation. So, the optimization of the scaffold's porosity is important to help cell growth, formation of vascularization, and the diffusion of nutrients [67].

Roohani-Esfahani and their co-workers developed a glass–ceramic scaffold, with dimension size 6x6x6mm, by direct ink writing mimicking cortical bone with 600 μm custom-made nozzle. In the work, they concluded that a scaffold with hexagonal pore shapes (450 μm, 550 μm, 900 μm, and 1200 μm) present the highest compressive strength, compared to the other designs [68].

Abbot and co-workers, in 2016, developed a silk scaffold with osteoblasts to evaluate in vitro culture that stimulated bone differentiation and regeneration. In the end, they concluded that it was evident the mineralization in the scaffold with silk seeded with this type of cells [23].

Tovar and co-workers [49] had developed a cylindrical scaffold with 10.5 mm length, 4.5 mm outside diameter and 2.25 mm inside diameter, 330 μm struts, and around 400 μm pore spacing. They used a 330-μm-diameter extrusion nozzle with a velocity of 8 mm/s. The existence of macrometric and micrometric porosity in the scaffold helped in its degradation, which allowed the biomechanical load to the healing bone. This may explain the rapid development of bone properties in the regenerated tissue that is highly indicative of complete healing when it is complemented with the remodeling of the original bone morphology.

Recently, Barba et al. [19, 69] concluded that the geometric parameters of the scaffold, like curvature, influence bone tissue regeneration. They demonstrated that spongy scaffolds with concave pores attracted a large amount of ectopic bone compared with scaffolds with prismatic geometries.

Through the existing studies, both in vivo and in vitro, one can get an idea of both material and biological components essentials to a proper bone regeneration. In the first phase, it is necessary to understand which scaffold design is the most appropriate and which biomaterials are to combine it with the AM technique.

5. Material component

For bone regeneration, it is necessary to consider that the scaffold must restore the normal biomechanical role of the tissue. **Table 2** shows different types of existing materials and their characteristics and some guidelines about how to obtain an ideal scaffold. However, there are other important features that need to be taken into account related to the different biological and physical signals involved in order to simulate the mechanism of remodeling in a natural environment, but more importantly, the scaffold must have the exact mechanical properties to withstand the loads the original bone held [6, 70]. The biomaterials used in the scaffolds must have a suitable rate of degradation in order to support bone regeneration. This rate of degradation depends on the corrosion resistance of the material used, which is affected by the chemical and physical characteristics of the scaffold [36].

Having into consideration the referred above and combining this information with the natural organization of bone (trabecular and cortical), the best strategy for BTE should pass by the use of collagen type I in the trabecular bone region and Hap in the cortical zone [80, 81]. Despite their advantages and the fact that they are already present in the bone native structure, their proper manipulation is only possible through their combination with synthetic polymers. The most suitable are poly(ε-caprolactone) (PCL) or polylactic acid (PLA) because they are both approved by the Food and Drug Administration. PCL is a stable, biocompatible, biodegradable polymer (from 12 to 48 months) and easy to handle to achieve the desired mechanical properties. Due to its low melting point (60°C) [82, 83], it can be easily combined with the collagen. PLA is a biocompatible polymer, more hydrophilic than PCL, and its handling is similar to the use of PCL. Hydrophilicity accelerates polymer degradation as it accelerates polymer and scaffold moisture [84]. However, it has a high melting point, which could be combined with Hap. With these materials, scaffolds can be produced with two methods: conventional and additive manufacturing (AM).

Scaffolds		**References**
Natural polymers	Biomaterials are widely used because of their biocompatibility, degradation, bioactivity, mechanical kinetics, tissue nonspecificity, and their intrinsic structural similarity to the extracellular matrix of native tissues. They also promote biological recognition, which can positively support cell adhesion and function	[71, 72]
Synthetic polymers	Easy to manipulate the properties of the material to achieve the appropriate mechanical behavior. At a microscale it presents the architecture, 3D composition, and active molecular reactive groups. In a macroscale, they have porosity, stiffness, and elasticity	[6]
Metals	Are used in long bones to better attach to the bones where there is minimal movement between the implant and the host tissue and provide physiological loading functionality to the implant site	[15, 73]
Ceramics	Have been used because of their ability to sustain compressive loads	[6, 74, 75]
Ideal scaffolds		
Should exhibit the adequate mechanical properties, pore size, and biological activity, serve as cell support, and guarantee new bone formation and thus the use of more than one material (a natural with a synthetic one)		[76–79]

Table 2.
Characteristics of the different materials used to produce a scaffold.

5.1 Additive manufacturing

Some studies have used conventional methods for producing scaffolds. However, these methods have no adequate control over pore size and design or interconnectivity [8, 85]. In order to address these problems, since the mid-1980s [86], a new manufacturing type of technology called AM has emerged. Its potential is enormous and overcomes the capabilities of the conventional technologies to produce scaffolds with a complex architecture and with the intention to achieve an appropriate mechanical response to the desired application [36].

Nowadays there are several approaches to AM for various applications. The main approaches are fused filament fabrication (FFF), three-dimensional printing (3DP), stereolithography (SLA), and selective laser sintering (SLS). Each process goes through several steps: (i) development of the 3D model through computer-aided design (CAD); (ii) the files are stored in standard triangular language (STL) format, which is a CAD file format that supports 3D printing and computer-aided manufacturing (CAM); and (iii) these files are inserted into the input devices to create 3D models in a layer-by-layer process [36]. In addition, there are still two processes where it uses the same principles of layer manufacturing: selective laser melting (SLM) [87–90] and electron beam melting (EBM) [91–93]. Both are used to produce metal scaffolds, although SLM can also process polymers and ceramics [3, 94, 95].

FFF, **Figure 1**, or melt-extrusion is an extrusion-based process and is the simplest 3D printing method (see **Table 3**) [36, 96]. Fine thermoplastic polymers in the form of filaments or granules are cast and extruded through a nozzle that allows

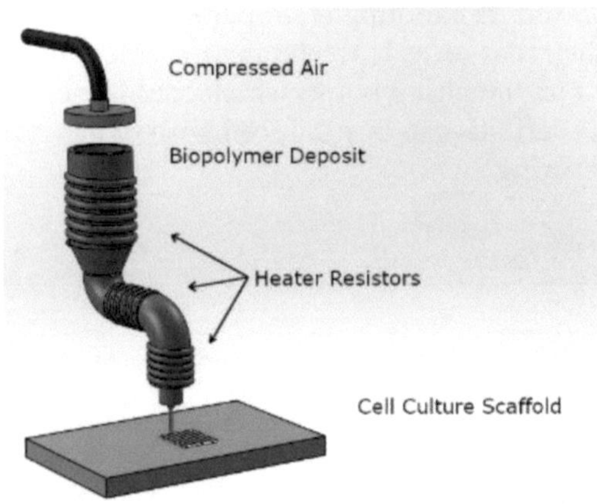

Figure 1.
Fused Filament Fabrication (FFF) process.

Fused filament fabrication	
Advantages	**Disadvantages**
Speed	Poor surface quality
Low cost	Need for heating in the molding process ➔ degradation of polymer materials
Simplicity	
Flexibility	

Table 3.
Advantages and disadvantages of the fused filament fabrication process.

flow in a horizontal and vertical plane (XY plane) [36]. To extrude it is necessary to have heating of the material, which causes degradation. However, the disadvantages of this technique can be overcome. The most suitable and desired mechanical properties can be achieved for the desired purpose with the combination of biomaterials. With this technique, it is already possible to extrude some bioceramics, such as HAp [97–99].

It is critical that the first layer is maintained at a temperature slightly below its set point to ensure successful adhesion between the layers. The 3D structure is determined by several factors, such as nozzle diameter, deposition rate, path spacing of the same layer, layer thickness, and deposition angle [96].

In the FFF technique, it is possible to control layer thickness and print orientation. The structural geometry of scaffolds is determined by the position and orientation of the filaments, which provide various pore shapes such as triangular, parallelogram, hexagonal, and also nonuniform shapes [100]. In this technique, there are two factors that affect the filament size, and consequently the pore size, which are the deposition velocity and the rotational velocity.

5.2 Temporary implants

It is necessary that the scaffolds in bone regeneration be biocompatible, biodegradable, osteoinductive (raising and cell maturation), and osteoconductive (provide a platform for cell growth) [39]. Scaffolds for bone regeneration should meet several specific criteria, such as filling any bone defect, ensuring pore interconnectivity, and having a pore architecture in order to promote bone formation and facilitate the exchange of oxygen bone growth [101–103]. The design of the scaffold can influence both the mechanical properties and cellular behavior [100, 104, 105] as highlighted in **Figure 2**.

A satisfactory bone growth leads to certain requirements. Porosity should be above 50% and pore size between 50 and 400 μm. It is difficult to achieve a "perfect" scaffold for bone regeneration due to pore design and size and a porosity distribution that mimics the native tissue [107, 108]. In the literature, there are no quantitative criteria that specify porosity or pore size or topology for bone regeneration. Porous scaffolds ranging in size from 50 to 500 μm are known to promote cell migration and vascularization, while micropores and nanopores control interaction with proteins and ion exchange with extracellular fluids [19, 109].

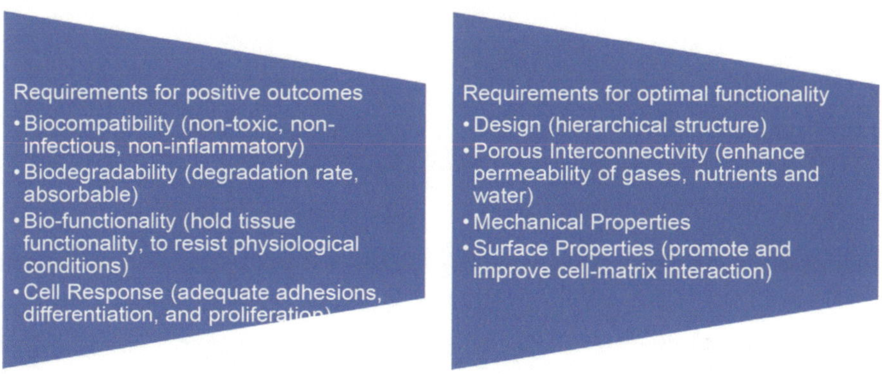

Figure 2.
Scaffold requirements in terms of response (left) and what should be taken into account (right) (adapted from [106]).

6. Biological components

The dogma of molecular biology is the basis for producing most bone cell and ECM components.

Bone remodeling is divided into five stages: activation, resorption, reversal, formation, and, finally, mineralization (see **Figure 3**). It is a process in which the old bone is reabsorbed and there is new bone formation. The cells that are involved in bone remodeling are osteoblasts, osteoclasts, and osteocytes, which actively participate in osseointegration and repair. Osteoclasts activate bone resorption, while osteocytes regulate bone homeostasis and osteoblasts form bone [15, 110].

In addition to bone cells, there are other cell lines that can be used in bone regeneration, which are human embryonic stem cells (hESCs), induced pluripotent stem cells (iPSCs), stem cells, and fibroblasts. According to Kuhn [111] and other workers [112, 113], hESCS present a rapidly proliferating rate. However, their transplantation induces uncontrollable spontaneous differentiation and can the teratoma formation may occur. Another type of stem cell is iPSCs. They can differentiate into several cells. However, there are studies that show that these cells can also give rise to teratomas and, in studies that distinguish high-quality lines from the iPSCs, allowed the detection of large duplications of genes that could potentially affect the differentiation and pluripotency of these cells [28, 99]. For these reasons, these cells are not considered the best ones for bone regeneration. Compared to fibroblasts, stem cells have a greater ability to migrate, so these type of cells are the most suitable cells for bone regeneration. Kargozar [58] recently studied the osteogenic potential of different MSCs, such as those derived from human bone marrow, umbilical cord (UC-MSCs) and adipose (AD-MSCs). It concluded that BMMSCs, according to collected histological data, is the most appropriate.

The combination of scaffold, AM, and bioreactor culture shows great potential for creating automated production ecosystems that will enable the formation of commercially available products for BTE application. Efficient nutrient and oxygen transport are important for this type of applications. To this end, bioreactor systems have tried to overcome this difficulty. Rotating-wall vessels are limited to small scaffolds as they do not provide optimal mass transport to the center of the scaffold and are not efficient in osteogenic differentiation, due to shear stress values transmitted to cells.

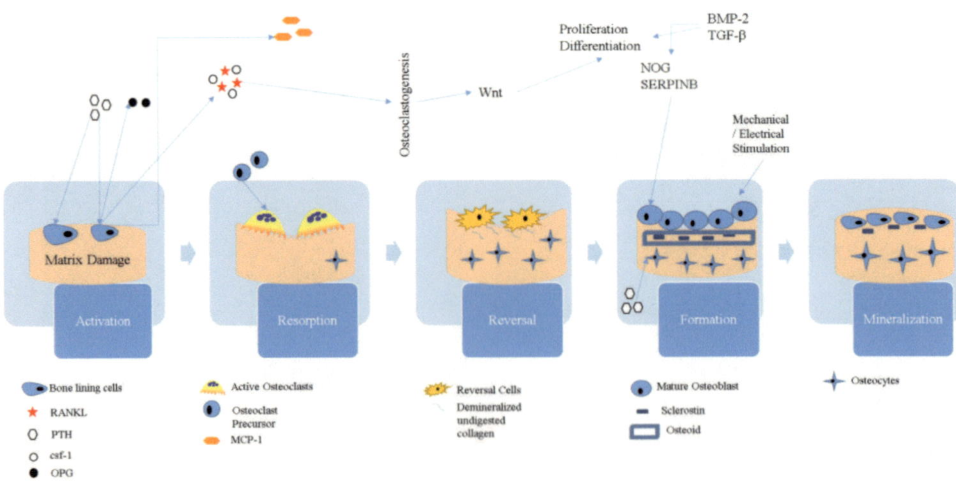

Figure 3.
Bone Remodeling Cycle.

On the other hand, agitated tanks have a major disadvantage regarding the circulating flow pattern that strikes cells against the bioreactor wall, which damages them and can lead to cellular apoptosis [114]. Finally, perfusion bioreactors are the best suited for BTE as they promote oxygenation throughout the whole scaffold, through improved mass transfer and shear stress, can expose cells to mechanical stimulation, and, therefore, obtain a much better cell distribution [6, 39, 61, 115, 116].

7. Optimized approach of a 3D scaffold

Scaffold architectures were designed in SolidWorks 2018 software. The design was bioinspired. This inspiration came from the natural organization of long bones, as represented in **Figure 4**. The diaphysis is composed of cortical bone (external region) which covers the trabecular bone (internal region). The trabecular bone has a larger surface area than the cortical bone and has a bone volume fraction ranging from 5% to a maximum of 60% [117]. It is known that the cortical zone corresponds to ~20% of the total diameter [118]. Bearing this in mind, it is expected that mimicking this type of organization, the mechanical behavior of the final scaffolds would be better and closer to the natural tissue.

The design considered has a height of 10 mm and diameter of 30 mm (see **Figure 5**). Thus, the cortical zone, the outer part of the scaffold, has a thickness of 6 mm and the trabecular zone, the inner part, has 18 mm. In the middle, there is a canal that corresponds to the medullar cavity. As happens in the native tissue, the region corresponding to the trabecular bone presents a higher porosity than the cortical one. So, the proposed scaffold has pores with different sizes between the different parts, bigger in the trabecular and smaller in the cortical

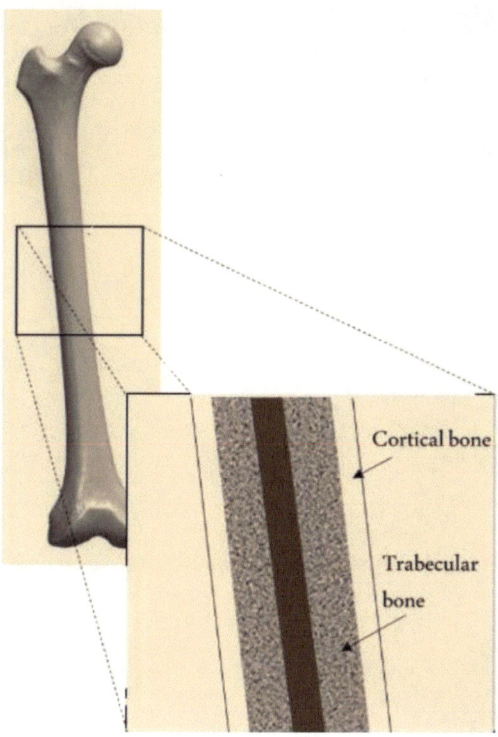

Figure 4.
Natural organization of long bones.

(400 and 300 μm of pore diameters, respectively). According to Zhang et al. [119], these pores are within the required values, since exceeding the pore size of 400 μm, cells do not sense the 3D, resulting in poor ECM production. Moreover, they are organized in a radial way, with a significant difference between the cortical zone and the trabecular zone.

The projected scaffold presents a total porosity of 42%, whereas the cortical part has approximately 5% porosity. This porosity mimics the normal porosity in the native bone (see **Table 1**) as shown by Fernandez-Yague et al. [120] and Wang et al. [3]. The trabecular zone has a porosity of approximately 57%. This porosity is also in agreement with the authors previously mentioned. However, this value is closer to the lower limit. This porosity can be improved by the addition of horizontal channels, but its inclusion would decrease the mechanical behavior of the proposed scaffold.

Since pore interconnectivity is considered by some authors a key point to cell migration and proliferation, another design is proposed, and (see **Figure 6**) it is inspired by a DNA strand. As the scaffold gains height, the base rotates, with a rotation angle of 36°. This was considered to guarantee that the end of the filaments was supported on all layers. Also, in this case, pores diverge gradually, so that the differences between cortical bone and trabecular bone can be noticed.

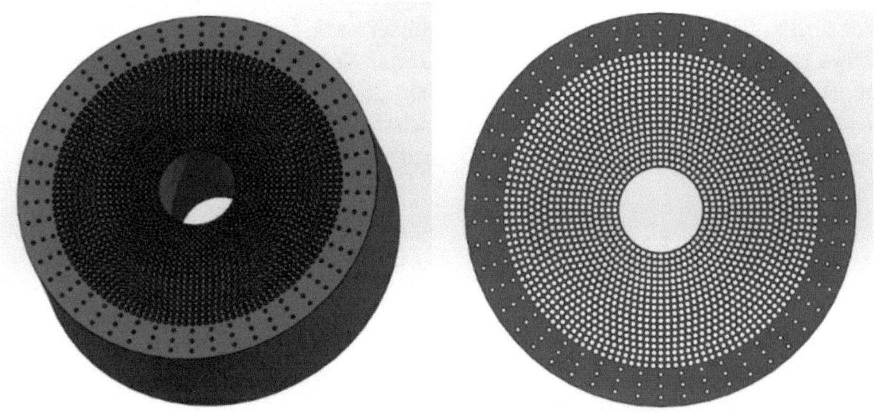

Figure 5.
Cylindrical scaffold.

Figure 6.
DNA chain-inspired cylindrical scaffold.

In this scaffold, pores range from 50 to 1500 μm in each layer, which is in agreement with the Zadpoor [121] and Szpalski [39]. However, the minimum required pore size is 100 μm, to make it easier to transport oxygen and nutrients and discard waste products [39, 121, 122]. For proper cell propagation, according to Bael [123], pore size should not pass the 1000 μm. Compared to the previous design presented, this one presents a slightly lower porosity in total, of around 38%, in which the cortical and trabecular parts have 22% and 49% porosity, respectively.

According to the authors Andrzejewska [34], Keaveny et al. [2], and Xiao et al. [124], the porosity of the trabecular zone is near the defined porosity values. However, the porosity of the zone corresponding to the cortical part is far above the maximum value of the defined values found in **Table 1**. In order to decrease this porosity, it is necessary to shrink the pore size used. Despite this limitation, this scaffold already has the advantage of fully interconnected pores, which will facilitate cell growth and the transport of oxygen and nutrients.

The combination of all supports of TE, which were described above, could lead successfully to bone formation. As biomechanics and TE advanced, it is easy to foresee the development of a new model for bone formation in which the use of an original scaffold leads to long bone fracture healing.

8. Conclusions

Bone defects are a constantly growing problem, affecting thousands of people around the world, which causes a loss in life quality, and most of the time, for an active population, it may take long periods of recovery. Until now, there are no synthetic substitutes that meet the mechanical and biological requirements for the long-term cure of critical-size bone defects. To overcome this health problem, the use of temporary biocompatible and biodegradable scaffolds becomes the best choice. Structures produced by AM have superior advantages compared to the conventional techniques, mainly due to better control over the desired architecture. Moreover, the choice of the AM technique to produce these scaffolds is essential to ensure control, namely, in terms of biological, physicochemical, and mechanical properties.

Considering all types of materials available, associated with the desired bone regeneration and the use of synthetic polymers, as PCL or PLA, combined with collagen type I for the trabecular region and Hap for cortical region, seems to be the best strategy to follow. To obtain the designed structures with these biomaterials, the most suitable AM technique is the FFF. For the selection of the final scaffold within the two proposals, further studies need to be performed. However, a third option could be also considered, which would include the cortical region of the first proposed scaffold (ensuring the required mechanical resistance) and the trabecular zone of the second one (assuring a proper porosity and pore interconnectivity to allow cell migration, nutrient, and oxygen exchange).

Among the most commonly used bioreactors for bone regeneration, perfusion bioreactors appear as the most suitable, because it improves osteogenic proliferation and differentiation due to improved mass transfer and adequate shear stress. When making a design proposal for bone regeneration, it is necessary to study the mechanical effects, such as stress and tension, and link them

Acknowledgements

This work is supported by the Fundação para a Ciência e Tecnologia (FCT) and Centro2020 through the Project references UID/Multi/04044/2019,

PAMI—ROTEIRO/0328/2013 (NO. 022158), and MATIS (CENTRO-01-0145-FEDER-000014-3362). It is also funded by the projects insitu.Biomas (POCI-01-0247-FEDER-017771), Bone2Move (PTDC/CVT-CVT/31146/2017), and Stimuli2BioScaffolds (PTDC/EME-SIS/32554/2017).

Conflict of interest

The authors declare no conflict of interest.

Appendices and nomenclature

AM	additive manufacturing
β-TCP	beta-tricalcium phosphate
BMMSCs	bone marrow mesenchymal stem cells
BTE	bone tissue engineering.
CAD	computer-aided design
EBM	electron beam melting
ECM	extracellular matrix
FDA	Food and Drug Administration
FFF	fused filament fabrication
GF	growth factors
Hap	hydroxyapatite
hBMSCs	human bone marrow-derived mesenchymal stem cells
hMSCs	human mesenchymal stem cells
hESCs	human embryonic stem cells
MSC	mesenchymal stem cells
PCL	poly(ε-caprolactone)
PLA	polylactic acid
rBMMSCs	rat bone marrow stromal cell
SLA	stereolithography
SLM	selective laser melting
SLS	selective laser sintering
TCP	tricalcium phosphate
TE	tissue engineering

Author details

Cristiana Fernandes[1], Carla Moura[1], Rita M.T. Ascenso[2], Sandra Amado[1], Nuno Alves[1] and Paula Pascoal-Faria[1*]

1 Centre for Rapid and Sustainable Product Development of the Polytechnic Institute of Leiria, Leiria, Portugal

2 Computer Science and Communication Research Centre, Polytechnic Institute of Leiria, Leiria, Portugal

*Address all correspondence to: paula.faria@ipleiria.pt

References

[1] Gabinete de Estratégia e Planeamento (GEP). Acidentes de Trabalho de 2017. Ministério do Trabalho, Solidariedade e Segurança Social (MTSSS). 2018. p. 228

[2] Keaveny TM, Morgan EF, Niebur GL, Yeh OC. Biomechanics of trabecular bone. Annual Review of Biomedical Engineering. 2001;**3**(1):307-333

[3] Wang X, Xu S, Zhou S, Xu W, Leary M, Choong P, et al. Topological design and additive manufacturing of porous metals for bone scaffolds and orthopaedic implants: A review. Biomaterials. 2016;**83**:127-141. Available from: https://linkinghub.elsevier.com/retrieve/pii/S0142961216000144

[4] Cole JH, van der Meulen MCH. Whole bone mechanics and bone quality. Clinical Orthopaedics and Related Research. 2011;**469**(8): 2139-2149

[5] Shrivats AR, McDermott MC, Hollinger JO. Bone tissue engineering: State of the union. Drug Discovery Today. 2014;**19**(6):781-786. DOI: 10.1016/j.drudis.2014.04.010

[6] Hasan A. Tissue Engineering for Artificial Organs: Regenerative Medicine, Smart Diagnostics and Personalized Medicine. Qatar University. Wiley-VCH Verlag; 2017. Available from: http://onlinelibrary.wiley.com/store/10.1002/9783527689934.ch17/asset/ ch17.pdf?v=1&t=j29txyw2&s=9855d273955de36128e55a5983693ca018f71ea5

[7] De Witte T-M, Fratila-Apachitei LE, Zadpoor AA, Peppas NA. Bone tissue engineering via growth factor delivery: From scaffolds to complex matrices. Regenerative Biomaterials. 2018;**5**(4):197-211. Available from: https://academic.oup.com/rb/article/5/4/197/5035357

[8] Orciani M, Fini M, Di Primio R, Mattioli-Belmonte M. Biofabrication and bone tissue regeneration: Cell source, approaches, and challenges. Frontiers in Bioengineering and Biotechnology. 2017;**5**:1-15. Available from: http://journal.frontiersin.org/article/10.3389/fbioe.2017.00017/full

[9] Jariwala SH, Lewis GS, Bushman ZJ, Adair JH, Donahue HJ. 3D printing of personalized artificial bone scaffolds. 3D Printing and Additive Manufacturing. 2015;**2**(2):56-64. DOI: 10.1089/3dp.2015.0001

[10] Poh PSP, Valainis D, Bhattacharya K, van Griensven M, Dondl P. Optimization of bone scaffold porosity distributions. Scientific Reports. 2019;**9**(1):9170. Available from: http://www.nature.com/articles/s41598-019-44872-2

[11] Schemitsch EH. Size matters: Defining critical in bone defect size! Journal of Orthopaedic Trauma. 2017;**31**(10):S20-S22. Available from: http://insights.ovid.com/crossref?an=00005131-201710005-00005

[12] Oryan A, Alidadi S, Moshiri A, Maffulli N. Bone regenerative medicine: Classic options, novel strategies, and future directions. Journal of Orthopaedic Surgery and Research. 2014;**9**(1):18. Available from: http://josr-online.biomedcentral.com/articles/10.1186/1749-799X-9-18

[13] Leach JK, Mooney DJ. Bone engineering by controlled delivery of osteoinductive molecules and cells. Expert Opinion on Biological Therapy. 2004;**4**(7):1015-1027. DOI: 10.1517/14712598.4.7.1015

[14] Gibbs DMR, Vaezi M, Yang S, Oreffo RO. Hope versus hype: What can additive manufacturing realistically offer trauma and orthopedic surgery?

Regenerative Medicine. 2014;**9**(4):535-549. DOI: 10.2217/rme.14.20

[15] Agarwal R, García AJ. Biomaterial strategies for engineering implants for enhanced osseointegration and bone repair. Advanced Drug Delivery Reviews. 2015;**94**:53-62. DOI: 10.1016/j.addr.2015.03.013

[16] Decoster TA, Gehlert RJ, Mikola EA, Pirela-cruz MA. Management of Posttraumatic Segmental Bone Defects. Journal of the American Academy of Orthopaedic Surgery. 2004;**12**(1):28-38

[17] Garbuz DS, Frcs C, Masri BA, Czitrom AA. Biology of allografting. The Orthopedic Clinics of North America. 1998;**29**(2):199-204

[18] Aghaloo TL, Hadaya D. Basic principles of bioengineering and regeneration. Oral and Maxillofacial Surgery Clinics. 2017;**29**(1):1-7. DOI: 10.1016/j.coms.2016.08.008

[19] Barba A, Maazouz Y, Diez-Escudero A, Rappe K, Espanol M, Montufar EB, et al. Osteogenesis by foamed and 3D-printed nanostructured calcium phosphate scaffolds: Effect of pore architecture. Acta Biomaterialia. 2018;**79**:135-147. DOI: 10.1016/j.actbio.2018.09.003

[20] Manassero M, Viateau V, Deschepper M, Oudina K, Logeart-Avramoglou D, Petite H, et al. Bone regeneration in sheep using acropora coral, a natural resorbable scaffold, and autologous mesenchymal stem cells. Tissue Engineering. Part A. 2013;**19**(13-14):1554-1563

[21] Langer R, Vacanti JP. Tissue engineering. Science (80-). 1993;**260**:920-926

[22] Loh QL, Choong C. Three-dimensional scaffolds for tissue engineering applications: Role of porosity and pore size. Tissue Engineering. Part B, Reviews. 2013;**19**(6):485-502

[23] Abbott RD, Kimmerling EP, Cairns DM, Kaplan DL. Silk as a biomaterial to support Long-term three-dimensional tissue cultures. ACS Applied Materials & Interfaces. 2016;**8**(34):21861-21868

[24] Abbott RD, Kaplan DL. Strategies for improving the physiological relevance of human engineered tissues. Trends in Biotechnology. 2015;**33**(7):401-407. DOI: 10.1016/j.tibtech.2015.04.003

[25] Giannoudis PV, Harwood PJ, Tosounidis T, Kanakaris NK. Restoration of long bone defects treated with the induced membrane technique: Protocol and outcomes. Injury. 2016;**47**:S53-S61. DOI: 10.1016/S0020-1383(16)30840-3

[26] Gugala Z, Gogolewski S. Regeneration of segmental Diaphyseal defects in sheep tibiae using Resorbable polymeric membranes: A preliminary study. Journal of Orthopaedic Trauma. 1999;**13**(3):187-195

[27] Key JA. The effect of a local calcium depot on osteogenesis and healing of fractures. The Journal of Bone and Joint Surgery. 1934;**16**(1):176-184

[28] Fliefel R, Ehrenfeld M, Otto S. Induced pluripotent stem cells (iPSCs) as a new source of bone in reconstructive surgery: A systematic review and meta-analysis of preclinical studies. Journal of Tissue Engineering and Regenerative Medicine. 2018;**12**(7):1780-1797

[29] Dimitriou R, Jones E, Mcgonagle D, Giannoudis PV. Bone regeneration: Current concepts and future directions. BMC Medicine. 2011;**9**(1):66

[30] Jeon OH, Panicker LM, Lu Q, Chae JJ, Feldman RA, Elisseeff JH. Human

iPSC-derived osteoblasts and osteoclasts together promote bone regeneration in 3D biomaterials. Scientific Reports. 2016;**6**:1-11. DOI: 10.1038/srep26761

[31] Lu H, Liu Y, Guo J, Wu H, Wang J, Wu G. Biomaterials with antibacterial and osteoinductive properties to repair infected bone defects. International Journal of Molecular Sciences. 2016;**17**(3):334

[32] Lu C, Chang Y, Lin S, Li K, Hu Y. Recent progresses in gene delivery-based bone tissue engineering. Biotechnology Advances. 2013;**31**(8):1695-1706. DOI: 10.1016/j.biotechadv.2013.08.015

[33] van Rijt S, Habibovic P. Enhancing regenerative approaches with nanoparticles. Journal of the Royal Society Interface. 2017;**14**(129):20170093

[34] Andrzejewska A. Biomechanical properties of 3D-printed bone models. In: Biosystems. Faculty of Mechanical Engineering, UTP University of Science and Technology in Bydgoszcz. Vol. 176. Poland: Elsevier; 2019. pp. 1-676. DOI: 10.1016/j. biosystems.2019.01.001

[35] Vogl F, Bernet B, Bolognesi D, Taylor WR. Towards assessing cortical bone porosity using low-frequency quantitative acoustics: A phantom-based study. PLoS One. 2017;**12**(9):e0182617. Available from: https://dx.plos.org/10.1371/journal.pone.0182617

[36] Yuan L, Ding S, Wen C. Additive manufacturing technology for porous metal implant applications and triple minimal surface structures: A review. Bioactive Materials. 2019;**4**(1):56-70. DOI: 10.1016/j.bioactmat.2018.12.003

[37] Mills LA, Simpson AHRW. In vivo models of bone repair. Journal of Bone and Joint Surgery. British Volume (London). 2012;**94-B**(7):865-874

[38] Clinton T, Lanyon LE. Regulation of bone formation by applied dynamic loads. The Journal of Bone and Joint Surgery. 1984;**66**:397-402

[39] Szpalski C, Sagebin F, Barbaro M, Warren SM. The influence of environmental factors on bone tissue engineering. Journal of Biomedical Materials Research Part B: Applied Biomaterials. 2013;**101**(4):663-675

[40] Cowin SC. Bone poroelasticity. Journal of Biomechanics. 1999;**32**(3):217-238

[41] Peric M, Dumic-Cule I, Grcevic D, Matijasic M, Verbanac D, Paul R, et al. The rational use of animal models in the evaluation of novel bone regenerative therapies. Bone. 2015;**70**:73-86. DOI: 10.1016/j.bone.2014.07.010

[42] Ho-Shui-Ling A, Bolander J, Rustom LE, Johnson AW, Luyten FP, Picart C. Bone regeneration strategies: Engineered scaffolds, bioactive molecules and stem cells current stage and future perspectives. Biomaterials. 2018;**180**:143-162. DOI: 10.1016/j.biomaterials.2018.07.017

[43] Lammens J, Maréchal M, Geris L, Van der Aa J, Van Hauwermeiren H, Luyten FP, et al. Warning about the use of critical-size defects for the translational study of bone repair: Analysis of a sheep Tibial model. Tissue Engineering. Part C, Methods. 2017;**23**(11):694-699

[44] Nather A, David V, Teng JWH, Lee CW, Pereira BP. Effect of autologous mesenchymal stem cells on biological healing of allografts in critical-sized tibial defects simulated in adult rabbits. Annals of the Academy of Medicine, Singapore. 2010;**39**(8):599

[45] Khojasteh A, Eslaminejad MB, Nazarian H, Morad G, Dashti SG, Behnia H, et al. Vertical bone augmentation with simultaneous implant placement using particulate

mineralized bone and mesenchymal stem cells: A preliminary study in rabbit. The Journal of Oral Implantology. 2013;**39**(1):3-13

[46] Lee JY, Choi MH, Shin EY, Kang YK. Autologous mesenchymal stem cells loaded in Gelfoam® for structural bone allograft healing in rabbits. Cell and Tissue Banking. 2011;**12**(4):299-309

[47] Jang DW, Franco RA, Sarkar SK, Lee BT. Fabrication of porous hydroxyapatite scaffolds as artificial bone preform and its biocompatibility evaluation. ASAIO Journal. 2014;**60**(2):216-223

[48] Chowdhary R, Halldin A, Jimbo R, Wennerberg A. Influence of micro threads alteration on osseointegration and primary stability of implants: An FEA and In vivo analysis in rabbits. Clinical Implant Dentistry and Related Research. 2015;**17**(3):562-569

[49] Tovar N, Witek L, Atria P, Sobieraj M, Bowers M, Lopez CD, et al. Form and functional repair of long bone using 3D-printed bioactive scaffolds. Journal of Tissue Engineering and Regenerative Medicine. 2018;**12**(9):1986-1999

[50] Saravanan S, Chawla A, Vairamani M, Sastry TP, Subramanian KS, Selvamurugan N. Scaffolds containing chitosan, gelatin and graphene oxide for bone tissue regeneration in vitro and in vivo. International Journal of Biological Macromolecules. 2017;**104**:1975-1985. DOI: 10.1016/j.ijbiomac.2017.01.034

[51] Smith JO, Tayton ER, Khan F, Aarvold A, Cook RB, Goodship A, et al. Large animal in vivo evaluation of a binary blend polymer scaffold for skeletal tissue-engineering strategies; translational issues. Journal of Tissue Engineering and Regenerative Medicine. 2014;**11**(4):1065-1076

[52] Bragdon CR, Burke D, Lowenstein JD, Connor DOO, Ramamurti B, Jasty M, et al. Differences in stiffness between a cementless and cancellous bone into varying amounts of the interface porous implant vivo in dogs due implant motion. Clinical Orthopaedics. 1996;**11**(8):945-951

[53] Ko J-Y, Park S, Im G-I. Osteogenesis from human induced pluripotent stem cells: An in vitro and in vivo comparison with mesenchymal stem cells. Stem Cells and Development. 2014;**23**(15):1788-1797

[54] Kim HJ, Park J-S. Usage of human mesenchymal stem cells in cell-based therapy: Advantages and disadvantages. Development and Reproduction. 2017;**21**(1):1-10. Available from: http://www.ksdb.org/archive/view_article?pid=dr-21-1-1

[55] Omole AE, Fakoya AOJ. Ten years of progress and promise of induced pluripotent stem cells: Historical origins, characteristics, mechanisms, limitations, and potential applications. PeerJ. 2018;**6**:e4370. Available from: https://peerj.com/articles/4370

[56] Zhang Z. Bone regeneration by stem cell and tissue engineering in oral and maxillofacial region. Frontiers of Medicine in China. 2011;**5**(4):401-413

[57] Hoch AI, Leach JK. Concise review: Optimizing expansion of bone marrow mesenchymal stem/stromal cells for clinical applications. Stem Cells Translational Medicine. 2014;**3**(5):643-652

[58] Kargozar S, Mozafari M, Hashemian SJ, Brouki Milan P, Hamzehlou S, Soleimani M, et al. Osteogenic potential of stem cells-seeded bioactive nanocomposite scaffolds: A comparative study between human mesenchymal stem cells derived from bone, umbilical cord Wharton's jelly, and adipose

tissue. Journal of Biomedical Materials Research Part B: Applied Biomaterials. 2018;**106**(1):61-72. DOI: 10.1002/jbm.b.33814

[59] Lv Q, Nair L, Laurencin CT. Fabrication, characterization, and in vitro evaluation of poly(lactic acid glycolic acid)/nano-hydroxyapatite composite microsphere-based scaffolds for bone tissue engineering in rotating bioreactors. Journal of Biomedical Materials Research Part A. 2009;**91**(3):679-691

[60] Hosseinpour S, Ghazizadeh Ahsaie M, Rezai Rad M, Baghani MT, MotamedianSR, KhojastehA. Application of selected scaffolds for bone tissue engineering: A systematic review. Oral and Maxillofacial Surgery. 2017;**21**(2):109-129

[61] Kleinhans C, Mohan RR, Vacun G, Schwarz T, Haller B, Sun Y, et al. A perfusion bioreactor system efficiently generates cell-loaded bone substitute materials for addressing critical size bone defects. Biotechnology Journal. 2015;**10**(11):1727-1738. DOI: 10.1002/biot.201400813

[62] Kim HJ, Kim UJ, Vunjak-Novakovic G, Min BH, Kaplan DL. Influence of macroporous protein scaffolds on bone tissue engineering from bone marrow stem cells. Biomaterials. 2005;**26**(21):4442-4452

[63] Robling AG, Castillo AB, Turner CH. Biomechanical and molecular regulation of bone remodeling. Annual Review of Biomedical Engineering. 2006;**8**:455-498

[64] Freshney IAN, Estate G, Freshney I. Application of cell cultures to toxicology. In: In Cell Culture Methods for In Vitro Toxicology. Dordrecht: Springer; 2001. pp. 9-26

[65] Meyer U, Wiesmann HP, Kruse-LöslerB, HandschelJ, StratmannU, Joos U. Strain-related bone remodeling in distraction osteogenesis of the mandible. Plastic and Reconstructive Surgery. 1999;**103**(3):800-807. Available from: https://insights.ovid.com/crossref?an=00006534-199903000-00005

[66] Univer- S. Fabrication of Porous Hydroxyapatite Scaffolds as Artificial Bone Preform and its Biocompatibility Evaluation. 2014. pp. 216-223

[67] Lu JX, Flautre B, Anselme K, Hardouin P, Gallur A, Descamps M, et al. Role of interconnections in porous bioceramics on bone recolonization in vitro and in vivo. Journal of Materials Science. Materials in Medicine. 1999;**10**(2):111-120

[68] Roohani-Esfahani SI, Newman P, Zreiqat H. Design and fabrication of 3D printed scaffolds with a mechanical strength comparable to cortical bone to repair large bone defects. Scientific Reports. 2016;**6**(February 2015):1-8. DOI: 10.1038/srep19468

[69] Barba A, Diez-Escudero A, Maazouz Y, Rappe K, Espanol M, Montufar EB, et al. Osteoinduction by foamed and 3D-printed calcium phosphate scaffolds: Effect of nanostructure and pore architecture. ACS Applied Materials & Interfaces. 2017;**9**(48):41722-41736

[70] Frohlich M, Grayson W, Wan L, Marolt D, Drobnic M, Vunjak-Novakovic G. Tissue engineered bone grafts: Biological requirements, tissue culture and clinical relevance. Current Stem Cell Research and Therapy. 2008;**3**(4):254-264. Available from: http://www.eurekaselect.com/openurl/content.php?genre=article&issn=1574-888X&volume=3&issue=4&spage=254

[71] Chen F-M, Liu X. Advancing biomaterials of human origin for tissue engineering. Progress in Polymer

Science. 2016;**53**:86-168. DOI: 10.1016/j.progpolymsci.2015.02.004

[72] Renth AN, Detamore MS. Leveraging "raw materials" as building blocks and bioactive signals in regenerative medicine. Tissue Engineering. Part B, Reviews. 2012;**18**(5):341-362. DOI: 10.1089/ten.teb.2012.0080

[73] Ramazanoglu M, Oshida Y. Osseointegration and bioscience of implant surfaces-current concepts at bone-implant interface. In: Turkyilmaz I, editor. Implant Dentistry-A Rapidly Evolving Practice. Rijeka: InTechOpen; 2011

[74] De Peppo GM, Marolt D. Modulating the biochemical and biophysical culture environment to enhance osteogenic differentiation and maturation of human pluripotent stem cell-derived mesenchymal progenitors. Stem Cell Research & Therapy. 2013;**4**(5):1-11

[75] Connecting Mechanics HR. Bone cell activities in the bone remodeling process: An integrated finite element modeling. Frontiers in Bioengineering and Biotechnology. 2014;**2**(April): 1-12. Available from: http://journal.frontiersin.org/article/10.3389/fbioe.2014.00006/abstract

[76] Chesnutt BM, Viano AM, Yuan Y, Yang Y, Guda T, Appleford MR, et al. Design and characterization of a novel chitosan/nanocrystalline calcium phosphate composite scaffold for bone regeneration. Journal of Biomedical Materials Research Part A. 2009;**88**(2):491-502

[77] Hennessy KM, Pollot BE, Clem WC, Phipps MC, Sawyer AA, Culpepper BK, et al. The effect of collagen I mimetic peptides on mesenchymal stem cell adhesion and differentiation, and on bone formation at hydroxyapatite surfaces. Biomaterials. 2009;**30**(10):1898-1909. DOI: 10.1016/j.biomaterials.2008.12.053

[78] Cao H, Kuboyama N. A biodegradable porous composite scaffold of PGA/β-TCP for bone tissue engineering. Bone. 2010;**46**(2):386-395. DOI: 10.1016/j.bone.2009.09.031

[79] Fu SZ, Ni PY, Wang BY, Chu BY, Zheng L, Luo F, et al. Injectable and thermo-sensitive PEG-PCL-PEG copolymer/collagen/n-HA hydrogel composite for guided bone regeneration. Biomaterials. 2012;**33**(19):4801-4809. DOI: 10.1016/j.biomaterials.2012.03.040

[80] Murphy CM, Schindeler A, Gleeson JP, Yu NYC, Cantrill LC, Mikulec K, et al. A collagen–hydroxyapatite scaffold allows for binding and co-delivery of recombinant bone morphogenetic proteins and bisphosphonates. Acta Biomaterialia. 2014;**10**(5):2250-2258. DOI: 10.1016/j.actbio.2014.01.016

[81] Wahl D, Czernuszka J. Collagen-hydroxyapatite composites for hard tissue repair. European Cells and Materials. 2006;**11**:43-56

[82] Pereira RF, Freitas D, Tojeira A, Almeida HA, Alves N, Bártolo PJ. Computer modelling and simulation of a bioreactor for tissue engineering. International Journal of Computer Integrated Manufacturing. 2014;**27**(10):946-959. DOI: 10.1080/0951192X.2013.812244

[83] Dorati R, DeTrizio A, Modena T, Conti B, Benazzo F, Gastaldi G, et al. Biodegradable scaffolds for bone regeneration combined with drug-delivery systems in osteomyelitis therapy. Pharmaceuticals. 2017;**10**(4):96. Available from: http://www.mdpi.com/1424-8247/10/4/96

[84] Yao Q, Cosme JGL, Xu T, Miszuk JM, Picciani PHS, Fong H, et al. Three dimensional electrospun

PCL/PLA blend nanofibrous scaffolds with significantly improved stem cells osteogenic differentiation and cranial bone formation. Biomaterials. 2017;**115**:115-127. DOI: 10.1016/j.biomaterials.2016.11.018

[85] Roseti L, Parisi V, Petretta M, Cavallo C, Desando G, Bartolotti I, et al. Scaffolds for bone tissue engineering: State of the art and new perspectives. Materials Science and Engineering: C. 2017;**78**:1246-1262. DOI: 10.1016/j.msec.2017.05.017

[86] Tang D, Tare RS, Yang LY, Williams DF, Ou KL, Oreffo ROC. Biofabrication of bone tissue: Approaches, challenges and translation for bone regeneration. Biomaterials. 2016;**83**:363-382. DOI: 10.1016/j.biomaterials.2016.01.024

[87] Van der Stok J, Van der Jagt OP, Amin Yavari S, De Haas MFP, Waarsing JH, Jahr H, et al. Selective laser melting-produced porous titanium scaffolds regenerate bone in critical size cortical bone defects. Journal of Orthopaedic Research. 2013;**31**(5):792-799. DOI: 10.1002/jor.22293

[88] Sing SL, Wang S, Agarwala S, Wiria FE, Ha TMH, Yeong WY. Fabrication of titanium based biphasic scaffold using selective laser melting and collagen immersion. International Journal of Bioprinting. 2017;**3**(1):1719-1722. Available from: http://www.sciencemag.org/content/299/5613/1719.abstract

[89] Guo Y, Wu J, Xie K, Tan J, Yang Y, Zhao S, et al. Study of bone regeneration and osteointegration effect of a novel selective laser-melted titanium-tantalum-niobium-zirconium alloy scaffold. ACS Biomaterials Science & Engineering. 2019;**5**(12):6463-6473. DOI: 10.1021/acsbiomaterials.9b00909

[90] Zhang B, Pei X, Zhou C, Fan Y, Jiang Q, Ronca A, et al. The biomimetic design and 3D printing of customized mechanical properties porous Ti_6Al_4V scaffold for load-bearing bone reconstruction. Materials and Design. 2018;**152**:30-39. DOI: 10.1016/j.matdes.2018.04.065

[91] Ataee A, Li Y, Fraser D, Song G, Wen C. Anisotropic Ti-6Al-4V gyroid scaffolds manufactured by electron beam melting (EBM) for bone implant applications. Materials and Design. 2018;**137**:345-354. DOI: 10.1016/j.matdes.2017.10.040

[92] Surmeneva MA, Surmenev RA, Chudinova EA, Koptioug A, Tkachev MS, Gorodzha SN, et al. Fabrication of multiple-layered gradient cellular metal scaffold via electron beam melting for segmental bone reconstruction. Materials and Design. 2017;**133**:195-204. DOI: 10.1016/j.matdes.2017.07.059

[93] Lv J, Xiu P, Tan J, Jia Z, Cai H, Liu Z. Enhanced angiogenesis and osteogenesis in critical bone defects by the controlled release of BMP-2 and VEGF: Implantation of electron beam melting-fabricated porous Ti_6Al_4V scaffolds incorporating growth factor-doped fibrin glue. Biomedical Materials [Internet]. 2015;**10**(3):035013. DOI: 10.1088/1748-6041/10/3/035013

[94] Ngo TD, Kashani A, Imbalzano G, Nguyen KTQ, Hui D. Additive manufacturing (3D printing): A review of materials, methods, applications and challenges. Composites Part B: Engineering. 2018;**143**(February):172-196. DOI: 10.1016/j.compositesb.2018.02.012

[95] Zadpoor AA, Malda J. Additive manufacturing of biomaterials, tissues, and organs. Annals of Biomedical Engineering. 2017;**45**(1):1-11. Available from: http://link.springer.com/10.1007/s10439-016-1719-y

[96] Ji K, Wang Y, Wei Q, Zhang K, Jiang A, Rao Y, et al. Application of

3D printing technology in bone tissue engineering. Bio-Design and Manufacturing. 2018;**1**(3):203-210. Available from: http://link.springer.com/10.1007/s42242-018-0021-2

[97] Rodriguez G, Dias J, D'Ávila MA, Bártolo P. Influence of hydroxyapatite on extruded 3D scaffolds. Procedia Engineering. 2013;**59**(November 2015):263-269. DOI: 10.1016/j.proeng.2013.05.120

[98] Morouço P, Biscaia S, Viana T, Franco M, Malça C, Mateus A, et al. Fabrication of poly(ε-caprolactone) scaffolds reinforced with cellulose nanofibers, with and without the addition of hydroxyapatite nanoparticles. BioMed Research International. 2016:1-10. Available from: https://www.hindawi.com/journals/bmri/2016/1596157/

[99] D'Amora U, Russo T, Gloria A, Rivieccio V, D'Antò V, Negri G, et al. 3D additive-manufactured nanocomposite magnetic scaffolds: Effect of the application mode of a time-dependent magnetic field on hMSCs behavior. Bioactive Materials. 2017;**2**(3):138-145

[100] Gleadall A, Visscher D, Yang J, Thomas D, Segal J. Review of additive manufactured tissue engineering scaffolds: Relationship between geometry and performance. Burns and Trauma. 2018;**6**(1):19. Available from: https://burnstrauma.biomedcentral.com/articles/10.1186/s41038-018-0121-4

[101] Hollister SJ, Murphy WL. Scaffold translation: Barriers between concept and clinic. Tissue Engineering. Part B, Reviews [Internet]. 2011;**17**(6):459-474. DOI: 10.1089/ten.teb.2011.0251

[102] Billström GH, Blom AW, Larsson S, Beswick AD. Application of scaffolds for bone regeneration strategies: Current trends and future directions. Injury [Internet]. 2013;**44**:S28-S33. DOI: 10.1016/S0020-1383(13)70007-X

[103] Wu S, Liu X, Yeung KWK, Liu C, Yang X. Biomimetic porous scaffolds for bone tissue engineering. Materials Science & Engineering R: Reports. 2014;**80**:1-36. DOI: 10.1016/j.mser.2014.04.001

[104] An J, Teoh JEM, Suntornnond R, Chua CK. Design and 3D printing of scaffolds and tissues. Engineering [Internet]. 2015;**1**(2):261-268. DOI: 10.15302/J-ENG-2015061

[105] Hollister S. Porous scaffold design for tissue engineering. Nature Materials. 2005;4(7):518. Available from: http://sfx.lib.umich.edu:9003/sfx_local?url_ver=Z39.88-2004&url_ctx_fmt=info:ofi/fmt:kev:mtx:ctx&rft_val_fmt=info:ofi/fmt:kev:mtx:journal&rft.atitle=Porous scaffold design for tissue engineering (vol 4, pg 518, 2005)&rft.auinit=S&rft.aulast=Hollister&rft

[106] Karioja H, Jantunent P. Infotech Oulu Annual Report 2016—Electronics Materials, Packaging and Reliability Techniques (EMPART). University Oulu; 2017. Available from: https://www.oulu.fi/infotech/annual_report/2016/empart [cited 21 June 2019]

[107] Arabnejad S, Johnston RB, Ann J, Singh B, Tanzer M, Pasini D. High-strength porous biomaterials for bone replacement: A strategy to assess the interplay between cell morphology, mechanical properties, bone ingrowth and manufacturing constraints. Acta Biomaterialia. 2016;**30**:345-356. DOI: 10.1016/j.actbio.2015.10.048

[108] Heinl P, Müller L, Körner C, Singer RF, Müller FA. Cellular Ti-6Al-4V structures with interconnected macro porosity for bone implants fabricated by selective electron beam melting. Acta Biomaterialia. 2008;**4**(5):1536-1544

[109] Hannink G, Arts JJC. Bioresorbability, porosity and mechanical strength of bone substitutes:

What is optimal for bone regeneration? Injury. 2011;**42**(Suppl. 2):S22-S25. DOI: 10.1016/j.injury.2011.06.008

[110] Dean DD, Oefinger J, Schwartz Z, Boyan BD, Lohmann CH, Bonewald LF. Implant surface characteristics modulate differentiation behavior of cells in the osteoblastic lineage. Advances in Dental Research. 2009;**13**(1):38-48

[111] Kuhn LT, Liu Y, Boyd NL, Dennis JE, Jiang X, Xin X, et al. Developmental-like bone regeneration by human embryonic stem cell-derived mesenchymal cells. Tissue Engineering. Part A. 2014;**20**(1-2):365-377. DOI: 10.1089/ten.tea.2013.0321

[112] Barberi T, Willis LM, Socci ND, Studer L. Derivation of multipotent mesenchymal precursors from human embryonic stem cells. PLoS Medicine. 2005;**2**(6):e161. Available from: https://dx.plos.org/10.1371/journal.pmed.0020161

[113] Trounson A. Human embryonic stem cells: Mother of all cell and tissue types. Reproductive Biomedicine Online. 2002;**4**(December):58-63. DOI: 10.1016/S1472-6483(12)60013-3

[114] Sladkova M, de Peppo G. Bioreactor systems for human bone tissue engineering. Processes. 2014;**2**(2):494-525. Available from: http://www.mdpi.com/2227-9717/2/2/494/

[115] Kim J, Kennedy K, Vunjak-Novakovic G. Bioreactors in regenerative medicine. In: Principles of Regenerative Medicine. Columbia University. New York, NY, United States: Elsevier; 2019. pp. 787-803. Available from: https://linkinghub.elsevier.com/retrieve/pii/B978012809880600045X

[116] Asadi-eydivand M, Solati-hashjin M, Farzad A, Osman NAA. Effect of technical parameters on porous structure and strength of 3D printed calcium sulfate prototypes. Robotics and Computer-Integrated Manufacturing. 2016;**37**:57-67. DOI: 10.1016/j.rcim.2015.06.005

[117] Drača N, Tikvica A, Vukičević S, Eljuga D, Semenski D, Brnčić M. Biomechanical properties of bones from rats treated with sevelamer. Collegium Antropologicum. 2011;**35**(2):557-563

[118] Long Y, Leslie WD, Luo Y. Study of DXA-derived lateral–medial cortical bone thickness in assessing hip fracture risk. Bone Reports. 2015;**2**:44-51. DOI: 10.1016/j.bonr.2015.02.003

[119] Zhang L, Yang G, Johnson BN, Jia X. Three-dimensional (3D) printed scaffold and material selection for bone repair. Acta Biomaterialia. 2019;**84**:16-33. DOI: 10.1016/j.actbio.2018.11.039

[120] Fernandez-Yague MA, Abbah SA, McNamara L, Zeugolis DI, Pandit A, Biggs MJ. Biomimetic approaches in bone tissue engineering: Integrating biological and physicomechanical strategies. Advanced Drug Delivery Reviews. 2015;**84**:1-29. DOI: 10.1016/j.addr.2014.09.005

[121] Zadpoor AA. Bone tissue regeneration: The role of scaffold geometry. Biomaterials Science 2015;3(2):231-245. Available from: http://xlink.rsc.org/?DOI=C4BM00291A

[122] Takahashi Y, Tabata Y. Effect of the fiber diameter and porosity of non-woven PET fabrics on the osteogenic differentiation of mesenchymal stem cells. Journal of Biomaterials Science. Polymer Edition. 2004;**15**(1):41-57. DOI: 10.1163/156856204322752228

[123] Van Bael S, Chai YC, Truscello S, Moesen M, Kerckhofs G, Van Oosterwyck H, et al. The effect of pore geometry on the in vitro biological behavior of human periosteum-derived

cells seeded on selective laser-melted Ti_6Al_4V bone scaffolds. Acta Biomaterialia. 2012;**8**(7):2824-2834. DOI: 10.1016/j.actbio.2012.04.001

[124] Xiao W, Wang Y, Pacios S, Li S, Graves DT. Cellular and molecular aspects of bone remodeling. In: Kantarci A, Will L, Yen S, editors. Tooth Movement Frontiers of Oral Biology. 2016. pp. 9-16. Available from: https://www.karger.com/Article/FullText/351895

Chapter 11

3D Printed Bioscaffolds for Developing Tissue-Engineered Constructs

Shiplu Roy Chowdhury, Yogeswaran Lokanathan, Law Jia Xian, Fauzi Mh Busra, Muhammad Dain Yazid, Nadiah Sulaiman, Gargy Lahiry and Md Enamul Hoque

Abstract

Tissue engineering techniques enable the fabrication of tissue substitutes integrating cells, biomaterials, and bioactive compounds to replace or repair damaged or diseased tissues. Despite the early success, current technology is unable to fabricate reproducible tissue-engineered constructs with the structural and functional similarity of the native tissue. The recent development of 3D printing technology empowers the opportunities of developing biofunctional complex tissue substitutes via layer-by-layer fabrication of cell(s), biomaterial(s), and bioactive compound(s) in precision. In this chapter, the current development of fabricating tissue-engineered constructs using 3D bioprinting technology for potential biomedical applications such as tissue replacement therapy, personalized therapy, and in vitro 3D modeling for drug discovery will be discussed. The current challenges, limitations, and role of stakeholders to grasp the future success also will be highlighted.

Keywords: 3D printing, scaffold, drug delivery, regenerative medicine, tissue engineering

1. Introduction

3D printing is a process whereby a real object is created starting with a virtual 3D digital model. It was first developed in 1986 by Hull and Lewis which is an improved stereolithography system using photochemical processes in which light causes chemical monomers to link together to form polymers and generate a solid object [1]. This technology is capable to fabricate a super complex geometry or features by accurately follow the computer-aided design (CAD) model. The fabrication requires appropriate materials that gradually released and overlapped in layer-by-layer fashion by 3D printer. The type of material chosen is crucial to ensure the printed object that can be used for further settings and applications. Various types of metals, polymers, ceramics, and composites such as polycaprolactone (PCL), polyethylene glycol (PEG), polylactic acid (PLA), acrylonitrile butadiene styrene (ABS) plastic, stainless steel, titanium, calcium phosphate, and silica can be used as starting materials in 3D printing [2–4].

Generally, there are four main applications of 3D printing in the medical field, which are as follows: (a) drug delivery, (b) surgical devices/implants, (c) operative planning, and (d) tissue engineering [5–12]. The 3D printing application for drug delivery is extensively used in the pharmaceutical industry to develop sustained release medication [5]. Modulation of the shell thickness as well as the shape of the 3D printed capsule allows precise control of the drug release rate [13]. 3D printing enables a fast and cost-effective way of fabricating personalized medical implants. The capability of producing custom implants gets rid of the need for adjustments during surgery that saves time as well as reduces the cost of operation and the risk of medical complications. This is particularly beneficial where metal implant interfaces with living bone and tissue. The electron-beam melting (EBM) and direct metal laser sintering (DMLS) technologies are both now used in the production of standard and customized implants. Surgical tools are generally designed to work with many patients. However, by fabricating patient-specific tools, it would decrease the risk of complications during surgery [13, 14]. Patient anatomy will be imaged using imager and transferred into the 3D design in CAD to create suitable tools that can be easily controlled during operation. In operative planning, the 3D printing also would provide surgeons with a visualization of the complex injuries. They can plan and strategize their work and choose specific tools required. Some of the common applications that require a 3D model are complex pelvic trauma [15], pediatric deformities [16], and osteotomies [17]. Furthermore, advances in 3D printing technology enable the possibilities of constructing living human tissues in the lab hoping to demonstrate structural and functional similarities as native tissue in the human body [12]. The biggest challenge is to construct thick tissue and to ensure the diffusion of oxygen and nutrients for cellular viability [14].

2. 3D bioprinting for designing bioscaffold

The conventional method to develop an engineered-tissue product involved the initial fabrication of specific native tissue design followed by the provision of cells and biomolecules. However, this approach could contribute to two major drawbacks including limitation in cell distribution and reduction in cell growth due to low nutrient concentration at the core area [10]. Very commonly used techniques for fabricating 3D scaffolds include freeze casting, solvent casting, gas foaming, and salt leaching [18]. The technology advancement in tissue engineering has been contributed to the current approach through computer-aided layered manufacturing technique, which is also known as 3D bioprinting. Briefly, the 3D bioprinting technology involves the combination of the primary ingredients known as "bio-ink" that functions as a biological framework and various types of cells with the presence of chemical factors, and biomolecules to form a solid and functional *in situ* 3D living structure [19].

There are four different techniques under 3D bioprinting including inkjet printing, extrusion-based methods, light-induced (photopolymerization) methods and particle fusion-based methods [7, 20–27]. The first three abovementioned techniques have been widely used to fabricate biomaterial designs [7].

The inkjet-based 3D bioprinting (**Figure 1**), first developed by Thomas Boland from Clemson University in 2003, is a low-cost manufacturing process that performs high-speed printing for 3D structure [21]. Besides, it provides high-resolution printing output up to 50 μm and widely proven to support cell viability and growth [22]. However, the main drawbacks are dealing with a low concentration of printing ink could hamper the reliability of cell encapsulation and significantly affect print fidelity [23]. Besides, this approach potentially could damage the printed

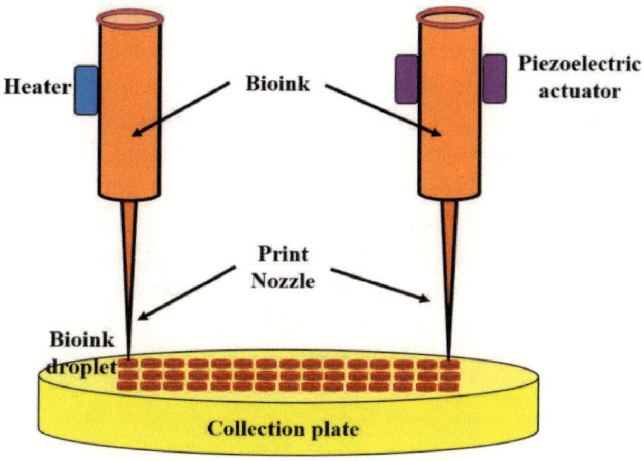

Figure 1.
The inkjet-based 3D bioprinting provides high resolution of printing output around 50 μm.

cells on the plantar under shear stress created through the inkjet-based printing but no concrete evidence is reported so far [21, 24]. Three main stages could affect the printable ink such as the production of the droplet, droplet/substrate close-interaction, and polymerization of the droplet. The two mechanisms, which have been involved under droplet generation through inkjet-based 3D bioprinting, are drop-on-demand and continuous inkjet [25]. The size of ink droplets produced via drop-on-demand and continuous injection is in the range of 25–50 and 100 μm, respectively [19]. Drop-on-demand inkjet has been conveniently used for tissue engineering applications.

The inkjet-based technology can be categorized into three as follows: thermal-based, piezoelectric-based, and magnetic-based inkjet printing [26]. The thermal induction can reach until 100–300°C that is required to nucleate a bubble and directly increase the appropriate pressure in the printhead lead to droplet expulsion [28]. There is no dead effect on the cells due to the presence of high temperature only for a microsecond and the previous study demonstrated consistency in cell viability post-inkjet-based 3D bioprinting [29]. Besides, the ink drop production can be induced by a piezoelectric method that focuses on the pulse pressure or acoustic waves generated from a piezoelectric actuator to expel printing ink drop. Another method to generate the drop expulsion is by using the electromagnetic approach depending on the Lorentz force and permanent magnet-based configurations. However, it produces a larger size of ink droplets as compared to thermal-based and piezoelectric-based approaches [28].

The second approach of 3D bioprinting is the extrusion-based method (**Figure 2**) that can be divided into two types consisting of fused deposition modeling and direct ink writing [19]. It is easy to handle, customized-based design bioprinter, and versatile with the developed current system. The principle of this 3D bioprinting method is that the printed ink extruded from the nozzle in liquid or molten state forms a particular line on the platform before polymerizing [30]. The bioprinting ink is commonly in the form of solid coil or filament that goes through the hot nozzle (temperature of around 200°C) before extrusion onto the platform. The extrusion from the printing nozzle is controlled by a specific system using various interventions including pressure-based control, pneumatic or mechanical control, or solenoid control before forming layered printed ink as required by the computerized set up to build up the 3D biomaterial designs [7]. The biopolymer should have an excellent solid-to-melt transition property to produce high-resolution 3D cell-laden on the printer platform [31]. However, the extrusion-based 3D bioprinting potentially could generate high mechanical force and shear stress together with high viscous of substrate lead to cellular apoptosis [5]. Further adjustment and optimization of this extrusion-based bioprinting can mitigate the drawbacks but it reduces the bioprinter resolution and speed [32]. Besides, the low concentration of ink viscosity supported cell proliferation and sustained the cell viability by introduced a composite-modified printing ink [33].

Light or laser-assisted 3D bioprinting, also known as stereolithography (SLA) (**Figure 3**), focuses on polymer resins manufacturing [19]. There are many variations of light or laser printing approaches for 3D fabrication. The advantages of these approaches are that they provide excellent accuracy, and good resolution between 10 and 50 μm [21]. This technique involves the patterning of a laser beam toward photo-based polymer to generate physical hardened polymer. This

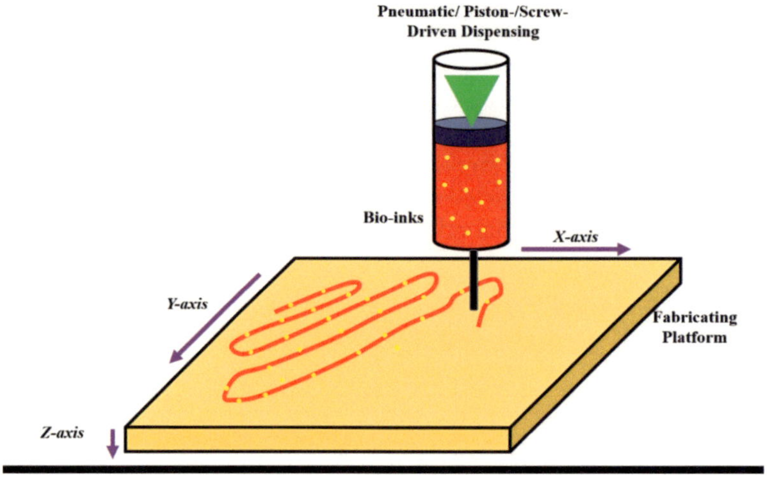

Figure 2.
The extrusion-based 3D bioprinting is easy to handle, customized-based design bioprinter, and versatile with the developed current system. It can be categorized into the fused deposition modeling and direct ink writing.

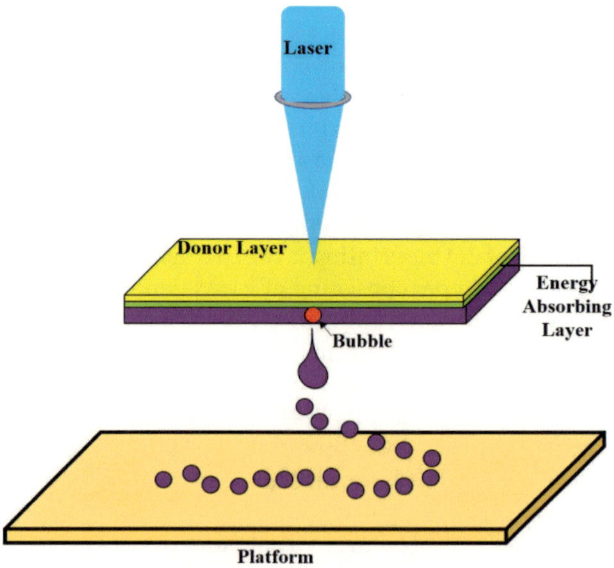

Figure 3.
Light- or laser-assisted 3D bioprinting approaches supported the high cell viability, accuracy, and good resolution between 10 and 50 μm. Two types consist of digital light processing-based bioprinting (DLP) and the two-photon polymerization-based bioprinting (TPP).

procedure is repeatedly applied to fabricate multi-layered polymer in the build-up stage. The other two types of laser-assisted 3D bioprinting techniques that are primarily applied in tissue engineering are digital light processing-based bioprinting (DLP) and the two-photon polymerization-based bioprinting (TPP) [34, 35]. The DLP technology uses a digitalized micro-mirror device chip (DMD) that contains around 2 million micro-mirrors. It functions to ensure light projection patterning precisely and is easy to modulate either on or off while the printing process is running on the platform. This technology consists of two 3D printing platform systems, namely, dynamic optical projection stereolithography (DOPS) and microscale continuous optical printing (μCOP) that support dynamic printing and continuous printing, respectively [7]. The TPP printing mechanism depends on the absorption of two photons by a molecule associated with light intensity square [36]. This phenomenon contributed to the printing of voxel dimension below 1 μm^3. Thus, this printing approach is an ideal method to generate nanoscale and microscale printing 3D output. However, high-resolution printing limits the construct size and printing speed. Nonetheless, the TPP printing speed is still faster than that in extrusion-based bioprinting and at a similar rate with inkjet-based bioprinting [37].

3. 3D bioprinting for developing tissue substitutes for therapeutic applications

Due to the limitation in technology to support the formation of the adequate and functional vascular network *in vitro*, currently, 3D bioprinting is more successful in the bioprinting of avascular tissue such as skin and cartilage. A complex tissue or organ with an extensive vascular network is still very challenging to prepare using the 3D bioprinting technology. To date, researchers are yet to succeed in preparing transplantable complex tissue or organ due to the difficulty in creating the circulatory system, especially the capillaries. However, several strategies have been used to improve the vascularization of 3D printed tissues, including printing of human umbilical vein endothelial cells (HUVECs) and vascular endothelial growth factors [38, 39] as well as seeding of endothelial cells and smooth muscle cells to the 3D printed tissues [40].

3.1 Bone

Bone tissue is one of the earliest tissues that were 3D printed and clinically used due to the ability of this technique to fabricate scaffolds according to the required shape, strength, and porosity. 3D printing enables fabrication of scaffold in any shape, which is not possible with many conventional fabrication techniques [41]. Furthermore, the materials commonly used for bone substitute production, such as hydroxylapatite (HA), synthetic calcium phosphate ceramics, polymethylmethacrylate, polylactides/polyglycolide and copolymer ceramics, tricalcium phosphate (TCP), bioglass, titanium, and other composite materials, are very compatible with the 3D printing technology [42]. The bone 3D printing had started as early as the 1990s, which utilized a powder-based freeform fabrication method [43]. Today, the bone substitute can be fabricated using the 3D plotting/direct ink writing, laser-assisted bioprinting (LAB), selective laser sintering (SLS), stereolithography (SLA), and fused deposition modeling (FDM) [42]. For example, Goriainov et al. prototyped hip joint implants using computer-aided design-computer-assisted manufacturing (CAD-CAM) and fabricated the scaffold using direct metal laser sintering from titanium alloy [44]. The custom-designed implants were seeded with autologous bone marrow aspirate before the implantation to 11 patients who were unsuitable for standard revision hip surgery. The postoperative results showed extensive new bone formation in the patients and a certain level of load-bearing function at the hip joint. The *in vitro* studies demonstrated the osteogenesis of the skeletal stem cells and osseointegration of the cells with the titanium alloy [44].

3.2 Skin

The other tissue that has a high potential to utilize 3D printing technology to repair and regenerate is skin. Although skin substitutes made by conventional tissue engineering techniques such as Matriderm®, Integra®, Dermagraft®, and OrCel® have been commercialized and been used in clinics for wound treatment, there are still challenges that are yet to be resolved by these skin substitutes. These skin substitutes are expensive, require long production time with prolonged healing time, have limited tissue functionality, and resulted in scarring in some cases [45]. Besides, these skin substitutes lack hairs, sweat glands, sebaceous glands, and other skin appendages as well as pigmentation. The 3D bioprinting technology has led to the paradigm shift in the skin substitute production where this transformative technology enables simultaneous and accurate deposition of multiple types of skin cells, the formation of scaffolds with complex macro- and micro-architecture, creation of vascular networks, and construction of stratified layer [46].

The commonly used skin 3D bioprinting techniques are microextrusion, inkjet, stereolithography, and laser-assisted bioprinting [47]. The materials commonly used in skin 3D bioprinting are mainly natural polymers such as alginate, gelatine, collagen, fibrin, and hyaluronic acid. However, biocompatible synthetic materials such as polycaprolactone (PCL), polyglycolide (PGA), polyethylene glycol (PEG), poly(lactic-co-glycolic acid) (PLGA), and polylactide (PLA) are commonly combined with natural polymers to increase the mechanical strength of the skin substitute [46, 47]. The bio-inks serve either as the cell carrier or sacrificial support that is removed after the printing, or both as a carrier and mechanical support material that provides greater strength and microarchitecture that supports the function of the skin even after the implantation on to the patients [46]. The on-site bioprinting of either autologous or allogeneic dermal fibroblasts and epidermal keratinocytes directly into a wound area is the latest development in skin 3D bioprinting. The direct deposition of the cells in fibrinogen/collagen solution in a layer-by-layer method onto porcine full-thickness wound has shown to promote the wound closure, reduce contraction, and enhance the re-epithelialization, and the regenerated skin tissue had the composition similar to healthy skin [48].

3.3 Vasculature

The other important and potential use of 3D bioprinting technology is the fabrication of vascularized tissues for passage of blood, air, lymph, and other vital fluids in the human body. The cells in dense tissue need to be within 200 mm from a vessel supplying oxygen and nutrients to survive [49]. The conventional technologies faced a major hindrance in fabricating vascular network structure in the dense engineered tissues, which is very crucial for the functioning of the implanted tissue or organ substitute, due to the technological limitation [50]. However, 3D bioprinting technology had enabled the fabrication of complex tissues with an integrated vasculature system, which in turn enabled the integration of the implant vasculature system with that of the host and long-term exchange of air, nutrient, and waste between the native and the implanted tissues [51].

The construction of the vasculature network throughout the tissue is achieved through the design and fabrication of the hollow tube structure in the micrometer scale. This hollow structure is also seeded with vascular cell types or angiogenic factors to promote the formation of functional microvascular networks structure, especially the branching that can size-up to the nanoscale range and also permeation capability [51]. The two main additive manufacturing concepts used for vascularized tissue formation are indirect and direct printing. In the indirect printing, a sacrificial material or negative mold is printed using thermo-reversible hydrogels such as Pluronic F-127 in combination with another material as the permanent scaffold. Upon completion of the 3D printing, the sacrificial mold is removed to form the vascular network that was cellularized with vascular cells [52]. In the direct printing method, the vascular structure is actively printed either with cell-loaded biomaterial or cell-compatible bio-ink. The bio-ink utilized in this process normally has quick gelation/cross-linking ability, or extrinsically induced to crosslink/cured, to maintain a stable hollow structure [52].

3D bioprinting has been utilized to prepare vascular networks in several studies. Miller et al. printed a 3D carbohydrate-glass lattice that was late embedded within an engineered tissue with living cells. Then, the 3D carbohydrate-glass lattice is removed, leaving interconnected hollow structures that can be seeded with endothelial cells to form the vasculature [53]. Later on, the same group of researchers proved that the vascular patch prepared using this technique can guide angiogenesis *in vivo* and rescue the ischemic tissues [54].

3.4 Other tissues

Besides the tissues discussed above, various other tissues have been and being fabricated with the still-evolving 3D bioprinting techniques. Many of these 3D printed tissues had also been implanted on patients as part of a clinical study [55–57] and systematic clinical trials are also being conducted for many of these products, which have been reviewed by Mehrotra et al. [58]. The 3D printed implants are in the clinical trial phase mostly as implants for an ankle injury, bone fracture, disease and deformation, and breast reconstruction. Among the other tissues that are in lab-scale fabrication and optimization but have a high potential for therapeutic use are liver tissue [59, 60], cardiac tissue [61, 62], kidney tissue [63, 64], pancreas tissue [65, 66], cartilage [67, 68], and neural tissues [69, 70].

Although the 3D bioprinting is a new technology, a few types of tissues produced by this technology are already utilized for therapeutic use. However, for the other tissues that have complex microarchitecture, and regulated by multiple signaling factors and cues from surrounding host tissues, it might need a longer time for the 3D printed tissue substitutes to be used in the clinical setting. The 3D printing of complex tissues needs more synergistic research from researches in various fields and various angles before it could fully mimic the native tissue's function. Another aspect to be considered will be the scaling up of the production using the clinical-grade materials and commercial-scale 3D printers as most of the current studies are being done with experimental materials and lab-scale 3D printer technologies.

4. 3D bioprinting for personalized therapy

Personalized medicine, also known as precision medicine, is a concept in medicine that emphasizes that each patient should be managed differently based on an individual's condition. This tailored therapy shall be able to provide the best treatment plan for the patients to improve their prognosis. In personalized medicine, all the patient's specific characteristics such as age, gender environment, height, weight, diet, environment, and genetics are being considered during the prevention, diagnosis, and treatment phase. Personalized medicine can improve the quality of patient care and reduce the cost by avoiding unnecessary diagnostic testing and treatments [71–73]. Personalized medicine is not only limited to drugs but also for tissue engineering and regenerative medicine. Tissue engineering is highly personalized as a specific tissue-engineered substitute is needed for each patient. For example, different burn patients are presented with different degrees of injury and varied wound location, size, and dimension. Thus, a unique engineered skin needs to be prepared in the current good manufacturing practice (cGMP) facility for each patient.

3D bioprinting is one of the techniques that allow the preparation of personalized tissue-engineered substitutes. One of the major advantages of 3D bioprinting in the field of tissue engineering is the possibility of producing personalized living tissue comprising of stem cells, cell-friendly matrix, and bioactive compound in the dimension uniquely suited for different patients. 3D bioprinting can be used to print simple living tissues like skin to a more complex hollow structure like a trachea and very complex organ like heart and kidney. This is something other living tissue fabrication techniques cannot achieve as these techniques do not allow precise deposition of cells at the space wanted. With the advances in the 3D bioprinting technology, nowadays, it is possible to print multiple types of cells, biomaterials, and bioactive compounds at different spaces to create a complex tissue that mimics the native tissue cellular arrangement and mechanical properties. Maturation of the 3D printed tissues can be achieved using a bioreactor.

To prepare the personalized 3D bioprinted living tissue, the image of the targeted tissue in specific patients needs to be taken and reconstructed into 3D, which will be used to guide the 3D printer to print the tissue in the dimension wanted layer-by-layer to form the 3D tissue [74]. Initially, 3D bioprinting is used to prepare engineered tissue *in vitro*, which can be transplanted *in vivo* afterward. However, it is difficult to maintain the shape and size of the engineered tissue *in vitro*. Thus, researchers come out with the idea of 3D bioprinting the tissue *in situ*, directly on the defect site (**Figure 4**). *In situ* 3D bioprinting allows the precise fitting of the printed tissue to the defect site, which is unique for every patient. *In situ* 3D bioprinting might be more efficient compared to the conventional technique as it allows more accurate reconstruction of defect sites and harnesses the natural healing capacity of the body to mature the printed tissue on time. An *in situ* 3D bioprinters can be as simple as a portable handheld spray gun to a complex robotic arm-assisted 3D bioprinter. Di Bella et al. developed an *in situ* handheld 3D bioprinter that printed mesenchymal stem cells encapsulated within the hyaluronic acid methacrylate-gelatin methacrylamide hydrogel surrounded by the hyaluronic acid methacrylate-gelatin methacrylamide hydrogel + photoinitiator VA-086 shell, which can be photocured using the ultraviolet right for the treatment of cartilage defect [75]. Keriquel et al. used 3D bioprinted mesenchymal stem cells in collagen with hydroxyapatite for bone tissue engineering in a mice model [76]. Cohen et al. used a robot-assisted method of *in situ* 3D bioprinting for the deposition of alginate hydrogel and demineralized bone matrix-gelatin hydrogel for the regeneration of cartilage and bone defects, respectively [77].

Apart from personalized engineered tissue substitutes, 3D bioprinting also can be utilized for the preparation of personalized drug delivery systems and functional tissue models for personalized drug screening and disease modeling. Various models have been developed, including the liver [78], heart [79], blood vessel [80], skin [81],

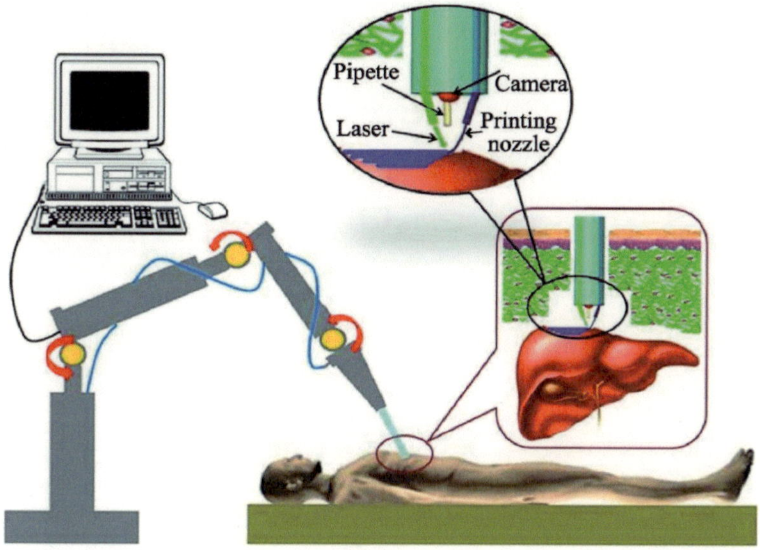

Figure 4.
The personalized 3D bioprinted living tissue has been printed layer-by-layer to form the 3D tissue.

skeletal muscle [82], and cancer [83]. The development of these models can greatly improve the medical care the patients will receive as distinctive prevention and treatment strategies can be designed individually.

5. 3D bioprinting for developing in vitro tissue/organoid models for drug discovery

The invention of 3D bioprinting has revolutionized biomedical research and significant development in translational research closing the gap from bench to bedside. In the pharmaceutical industry, the value of 3D bioprinting is expected in lowering the attrition rate of a new drug since 3D bioprinting has the potential to precisely position multiple cell types as needed according to the tissue of interest (**Figure 5**). Thus, 3D bioprinting enables a more robust design of drug screening, drug delivery, high-throughput drug testing, and ADME assays. The application of 3D bioprinting in the development of in vitro tissue or organoid models for drug discovery is discussed in this section.

5.1 Tumor or cancer model

The ability of 3D bioprinting in replicating tumor microenvironment (TME) provides a better model to assess drug response, tumor proliferation, and metastasis. By 3D bioprinting, a tumor model with hypoxic core and necrosis could be recreated similar to the *in vivo* environment [84, 85]. The 3D-printed glioma model comprising of glioma stem cells incorporated in alginate/gelatin/fibrinogen bioink is an example, and it showed higher resistance to temozolomide than in a 2D culture model [86]. Another case in point, fabrication of breast cancer model was achieved via the Organovo 3D NoveGen Bioprinter system where cancer cells are bordered with a stromal milieu of endothelial cells, fibroblast, and adipocytes. The said breast cancer model was viable for up to 14 days and possesses distinct internal compartmentalization. The model has been used to test hormonal drug response

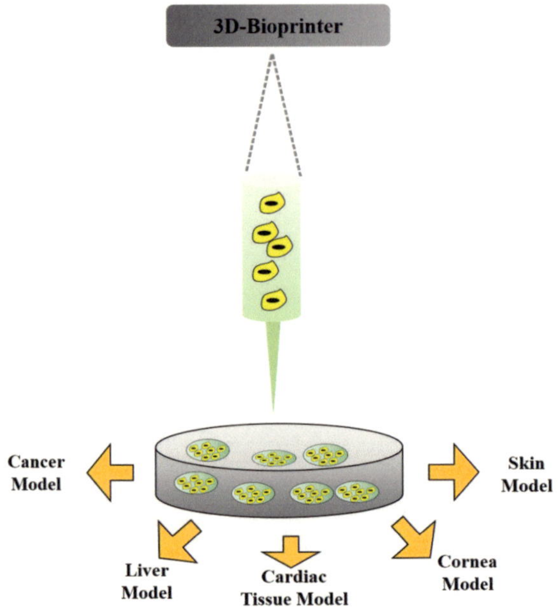

Figure 5.
The potential development of organoid models for drug discovery such as for cancer model, skin, cornea, intestines, muscle, cardiac tissue, and liver.

and chemotherapeutic agents [87]. Most reports conclude that 3D bioprinting gave a higher effect be it tolerance or resistance to the drug tested as compared to the 2D model of the disease, thus proving the value of 3D bioprinting in cancer drug screening.

5.2 Skin

The human skin's inherent multi-layered, multicellular composition is in demand commercially for pharmaceutical and dermatological testing. Dermal skin equivalent has been successfully created using 3D bioprinting through several approaches. One of them is via direct cell printing of fibroblasts and keratinocytes in the collagen-based hydrogel to recreate the skin stratification [88]. The incorporation of melanocytes and fibroblasts in collagen/fibroblast bio-ink was also reported [89]. Maturation and stratification of 3D bioprinted skin construct could be achieved via exposure to the air-liquid interface as shown by Lee et al. with skin construct expressing skin-specific markers [90]. These skin-like constructs are of value in drug toxicity screening as shown by Tseng et al. where five different drugs, i.e. all-trans retinoic acid, dexamethasone, doxorubicin, 5′-fluorouracil, and forskolin, use their 3D bioprinted fibroblasts [91].

5.3 Cornea

Corneal in vitro/ex vivo model is desperately needed as cornea function as major barrier in penetration of drugs into eye; thus, drug absorption thru cornea need to be optimized for topical ocular drug application. Hence, many studies were done in an animal model which is not cost-effective. The complex arrangement of collagen lamellae could be recapitulated using a 3D bioprinting system. Such a corneal model has been successfully produced utilizing extrusion-based bioprinting (EBB) of collagen/alginate/keratinocyte bio-ink [92]. Similar studies utilizing 3D bioprinting with a promising outcome have also been reported such as the generation of 3D multi-lamellar silk film incorporated with human corneal stromal stem cells (hCSSCs). The silk film architecture supports the growth and differentiation of hCSSCs in producing matured corneal stroma with the desired optical and mechanical properties close to the native cornea [93].

5.4 Intestines

Drugs are commonly absorbed in the intestine; hence, an in vitro intestinal tissue model is of value in the early phase of drug screening. Such a model was fabricated successfully using the Organovo 3D NovoGen bioprinter system with epithelial cells and myofibroblast that has a polarized columnar epithelium with tight junctions and specialized cells that express cytochromes P450 (CYP450). The above said model is a good model for Crohn's disease and internal bowel disease (IBD) that could be used in early-phase drug screening or toxicology study [94].

5.5 Muscle

Development of drugs that are delivered through intramuscular injection or for muscle injuries and muscular dystrophy require an in vitro muscle model for screening and testing. Alginate and Pluronic mixed with murine C2C12 cells have been successfully printed using the EBB method to create a 3D muscle construct that is used to screen several drugs and observe the cell viability, myogenic differentiation, and tissue contractile force against the drug [95, 96].

5.6 Cardiac tissue

Cardiovascular disease (CVD) is the leading cause of death in the world. Cardiotoxicity is the primary cause of CVD drug retraction from the market and is often done in 2D cell cultures. Therefore, the development of cardiovascular disease modeling and drug screening platform is a necessity. Most work focuses on recreating the left ventricular myocardium where cardiac pathologies occur. A spontaneously and synchronously contracting tissue was successfully developed with aligning endothelial cells that are used for cardiotoxicity screening [97]. In another study, Lind et al. fabricated self-assembled rat-derived cardiac cells by direct printing of six functional bio-inks that are highly conductance, piezoresistive, and biocompatible material. This model exhibits inotropic responses similar to isolated post-natal whole rat heart to several CVD drugs, i.e., L-type calcium channel blocker, verapamil, and β adrenergic agonist isoproterenol [98].

5.7 Liver

3D bioprinting approaches have been utilized in creating a liver disease model and liver tissue. Hepatotoxicity study of any drug introduced in the market is essential in any preclinical drug development. The establishment of in vitro liver models includes the incorporation of primary hepatocytes, hepatic cell lines, and stem cell-derived hepatic cells [99–101]. Kang et al. created a five-layer 3D hepatic structure using alginate and mouse induced hepatocyte-like cells that express albumin, ASGR1, and HNF4a [102]. Biomimetic liver tissue builds by Ma et al. showed better liver-specific function and drug metabolism potential compared to 2D monolayer culture [103].

The application of 3D bioprinting technology in the development of in vitro tissue or organoid models for drug discovery has fruitfully shown a better model in mitigating the risk associated with drug development. A 3D environment provides a better representation of an in vivo model in addition to reducing or eliminating the use of animal model early in the drug development process. All in all, the reliable prediction of safety and efficacy means a significant reduction of time and financial investment of a particular drug in question.

6. Challenges of 3D printing in tissue engineering

Although tissue engineering emerged with this glory for a few decades, the initial attempts took way long [104], whereas 3D printing of complex biomaterials is a promising means of scaffold designing.

There are different types of 3D printers: laser-, inkjet-, and extrusion-based. However, inkjet-based is more popular in tissue engineering, where cells or biomaterials are incorporated into the substrate, as per digitally set instruction, to recreate a functional organ or tissue. Multiple printheads can be used in the case of organs/tissue containing different types of cells. However, there are several challenges to address while designing a 3D printed engineered tissue [105].

6.1 Materials

6.1.1 Choice and processability of materials

The form of material input is important for this specialized process of 3D printing. Hence, it is important to think through before choosing a material, whether it is compatible to form a filament or powder or pellet or solution, that is required for that process. Another important feature to be considered while choosing the material is the expected mechanical strength of the scaffold and their biocompatibility and biodegradability.

6.1.2 Rate of biodegradation

The sole intent of engineered tissue is to replace and regenerate damaged tissue or organ. To comply with this requirement, the scaffold material of the transplanted tissue should be subject to remodeling and absorption. They should be able to degrade in equal or similar pace with the regeneration of extracellular matrix and differentiation of cells. This phenomenon depends on several factors, including hydrophilicity of the scaffold, surface area, porosity, degree of crystallinity, presence or absence of certain enzymes, etc. The most critical part here is harmonization in these factors, so that the degradation of biomaterial and stress release to the surrounding tissue is well synchronized, to ensure healing of the damaged tissue.

6.1.3 Biodegradation of product

Biodegradation rate affects the cell viability and mobility, despite the general concept of this biodegradation being non-cytotoxic. The study finds that the fast degradation of the polymer may affect the cells negatively due to the formation of acidic byproduct. However, more research is required to support these data and to develop the degradation profile of the materials.

6.1.4 Mechanical strength

Cells are described to be sensitive toward the mechanical strength of the polymer scaffold. Rigid and non-flexible material may hinder the cytoskeleton assembly, cell organization, and receptor recruitment into "focal adhesion plaques," which is crucial for cell signaling and anchoring. On the other hand, highly pliable material may not be able to provide the mechanical strength for anchoring or cytoskeleton assembly and thus affecting the cellular function as well.

6.2 Designing the polymer scaffold structure

6.2.1 Porosity

Different tissues require different porosities for the optimum effect. However, little knowledge is available. A general range of pore size is suggested to be considered for any type of cell, based on observations, rather than the established theory of optimum pore size for each cell type.

6.2.2 Morphology of the polymer scaffold

A study by Yin et al. describes that the microgrooves on the scaffold surface directly affect the cardiac function and susceptibility to arrhythmias [106]. This indicates the importance of the scaffold surface microenvironment, which positively or negatively affects the success of the tissue transplant.

6.2.3 Surface topography

It is stated that surface roughness may enhance adhesion between cell and extracellular matrix. At the same time, too rough surface of the scaffold may exhaust the cell adhesion capability. On the other side, if the scaffold material is too sharp, the cells may get damaged. However, choosing a smooth surfaced scaffold material may require consideration of further modification or coating, as this feature does not facilitate cell adhesion.

6.3 Vascularization

Small and simpler organ printing has been successful, without much difficulty. However, it is not simple when comes to bigger and complex organ, due to difficulty in vascularization. Small tissues are avascular, and most of the time, aneural, alymphatic, and thin or hollow. They can receive nutrition from host vasculature. But when the transplanted tissue is thicker than 150–200 μm, oxygen cannot be diffused from host tissue to it. As such, to create a functional bigger and complex tissue or organ, an integrated vascular system is to be created, which is still not in place [105].

6.4 Cell seeding

The homogenous distribution of the cells throughout the scaffold is important for the effectivity of the tissue. The conventional usage of Petri dish may not be adequate to ensure the uniform seeding of these cells. The bioreactor technology can influence a successful cell seeding, throughout the depth of the scaffold, evenly.

6.5 Future prospect of 3D printing

Despite all the challenges, 3D bioprinting offers great potential and diverse applications for the medical and healthcare sector.

6.5.1 Complex organ engineering

Although few technical aspects are still to be figured out, rapid prototyping creates possibilities to generate complex organs like kidney, liver or even heart, despite having a heterogeneous cellular composition. With the fast pace of advancement in technology and the number of researches going on in this field, the current challenges are expected to be resolved eventually. It has been foreseen that within 20 years, 3D printed organs will be commercially available for transplantation [105].

6.5.2 In vivo test models

The animal study is a mandatory part of drug designing, which applies to tissue engineering and cell therapy. It has been estimated that about 115 million animals are being used in the biomedical industry per year [107]. The printed organs can replace these animal tests of safety, efficacy, and toxicology, saving a number of animals, and resolving the ethical conflict in this issue. At the same time, these printed organs can be more "close to the human subject" model than the animals.

6.5.3 New drug design

In an optimistic vision, it may be possible to have a printed piece of patient's tissue to test which drug is suitable and effective for that particular patient, before applying on them, using this technology.

6.5.4 Mass production

Conventional tissue engineering involves customized scaffold preparation and manual cell seeding. Hence, the success rate is not consistent and the production cost is high, thereby resulting in very costly tissue that many people cannot afford. With the automation and advancement of bioreactor technique in conjunction with rapid prototyping, mass production of the complete organ is a very likely prospect [104]. This will increase the efficiency of the procedure of organ formation, and mass production capability will be economic and more affordable.

6.5.5 Less dependency on organ donation

The organ donation rate has always been far less than the requirement in a given period. On top of that, immunogenicity, rejection, and graft-versus-host disease make the transplantation process further difficult. With rapid prototyping, the scarcity of human organs can be resolved, with less immune rejection and higher effectivity.

6.5.6 In situ tissue printing

In situ generation of skin has already been achieved. With the progress of this technology, it is deemed that in future, a small piece of any tissue can be bioprinted in situ, during surgery, in no time, with precision [105].

7. Conclusions

Regenerative medicine is the new big thing in the medical and healthcare areas. Due to the promising outcome and compatibility for the human body, this alternative treatment method might compete and/or take over conventional medicine soon. Rapid prototyping has a very wide prospect in regenerative medicine, medical device, and pharmaceuticals. Incorporating the knowledge of cellular biology, biomaterial design, tissue engineering, bioreactors, and so on, organ regeneration will be much more precise and effective. With this speed of progression of science, the remaining challenges will be resolved soon, thereby opening a new era of healthcare and a better life for human beings.

Conflict of interest

The authors declare no conflict of interest.

Author details

Shiplu Roy Chowdhury[1], Yogeswaran Lokanathan[1], Law Jia Xian[1], Fauzi Mh Busra[1], Muhammad Dain Yazid[1], Nadiah Sulaiman[1], Gargy Lahiry[2] and Md Enamul Hoque[3]*

1 Tissue Engineering Centre, Universiti Kebangsaan Malaysia Medical Centre, Kuala Lumpur, Malaysia

2 Yakin Medic Sdn Bhd, Kuala Lumpur, Malaysia

3 Department of Biomedical Engineering, Military Institute of Science and Technology, Dhaka, Bangladesh

*Address all correspondence to: enamul1973@gmail.com

References

[1] Hull CW, Arcadia C. Apparatus for Production of Three-Dimensional Objects by Stereolithography. US Patent 4575330; 1986

[2] Hoque ME, Hutmacher DW, Feng W, Li S, Huang MH, Vert M, et al. Fabrication using a rapid prototyping system and in vitro characterization of PEG-PCL-PLA scaffolds for tissue engineering. Journal of Biomaterials Science. Polymer Edition. 2005;**16**(12):1595-1610

[3] Hoque ME, San WY, Wei F, Li S, Huang MH, Vert M, et al. Processing of polycaprolactone and polycaprolactone-based copolymers into 3D scaffolds, and their cellular responses. Tissue Engineering. Part A. 2009;**15**(10):3013-3024

[4] Hoque ME, Meng TTH, Chuan YL, Chowdhury M, Prasad RGSV. Fabrication and characterization of hybrid PCL/PEG 3D scaffolds for potential tissue engineering applications. Materials Letters. 2014;**131**:255-258

[5] Gioumouxouzis CI, Karavasili C, Fatouros DG. Recent advances in pharmaceutical dosage forms and devices using additive manufacturing technologies. Drug Discovery Today. 2019;**24**(2):636-643

[6] Peng W, Datta P, Ayan B, Ozbolat V, Sosnoski D, Ozbolat IT. 3D bioprinting for drug discovery and development in pharmaceutics. Acta Biomaterialia. 2017;**57**:26-46

[7] Ma X, Liu J, Zhu W, Tang M, Lawrence N, Yu C, et al. 3D bioprinting of functional tissue models for personalized drug screening and in vitro disease modeling. Advanced Drug Delivery Reviews. 2018;**132**:235-251

[8] Hoque ME. Robust formulation for the design of tissue engineering scaffolds: A comprehensive study on structural anisotropy, viscoelasticity and degradation of 3D scaffolds fabricated with customized desktop robot based rapid prototyping (DRBRP) system. Materials Science & Engineering. C, Materials for Biological Applications. 2017;**72**:433-443

[9] Hoque ME, Feng W, Wong YS, Hutmacher DW, Li S, Huang MH, et al. Scaffolds designed and fabricated with elastic biomaterials applying CAD-CAM technique. Tissue Engineering Parts A. 2008;**14**:907

[10] Derakhshanfar S, Mbeleck R, Xu K, Zhang X, Zhong W, Xing M. 3D bioprinting for biomedical devices and tissue engineering: A review of recent trends and advances. Bioactive Materials. 2018;**3**(2):144-156

[11] Dang TT, Nikkhah M, Memic A, Khademhosseini A. Polymeric biomaterials for implantable prostheses. In: Natural and Synthetic Biomedical Polymers. UK: Elsevier; 2014

[12] Hoang D, Perrault D, Stevanovic M, Ghiassi A. Surgical applications of three-dimensional printing: A review of the current literature & how to get started. Annals of Translational Medicine. 2016;**4**(23):456

[13] Okwuosa TC, Soares C, Gollwitzer V, Habashy R, Timmins P, Alhnan MA. On demand manufacturing of patient-specific liquid capsules via co-ordinated 3D printing and liquid dispensing. European Journal of Pharmaceutical Sciences. 2018;**118**:134-143

[14] AlAli AB, Griffin MF, Butler PE. Three-dimensional printing surgical applications. Eplasty. 2015;**15**:e37

[15] Falchi M, Rollandi GA. CT of pelvic fractures. European Journal of Radiology. 2004;**50**(1):96-105

[16] Cheung CL, Looi T, Lendvay TS, Drake JM, Farhat WA. Use of 3-dimensional printing technology and silicone modeling in surgical simulation: Development and face validation in pediatric laparoscopic pyeloplasty. Journal of Surgical Education. 2014;**71**(5):762-767

[17] Cherkasskiy L, Caffrey JP, Szewczyk AF, Cory E, Bomar JD, Farnsworth CL, et al. Patient-specific 3D models aid planning for triplane proximal femoral osteotomy in slipped capital femoral epiphysis. Journal of Children's Orthopaedics. 2017;**11**(2):147-153

[18] Zhang LG, Fisher JP, Leong KW, Starly B, Shirwaiker R. 3D bioprinting techniques. In: 3D Bioprinting and Nanotechnology in Tissue Engineering and Regenerative Medicine. UK: Elsevier; 2015

[19] Guvendiren M, Molde J, Soares RMD, Kohn J. Designing biomaterials for 3D printing. ACS Biomaterials Science & Engineering. 2016;**2**(10):1679-1693

[20] Hutmacher DW, Hoque ME, Wong YS. Design, fabrication and physical characterization of scaffolds made from biodegradable synthetic polymers in combination with RP systems based on melt extrusion. In: Bidanda B, Bártolo P, editors. Virtual Prototyping & Bio Manufacturing in Medical Applications. USA: Springer Science+Business Media, LLC; 2008

[21] Cidonio G, Glinka M, Dawson JI, Oreffo ROC. The cell in the ink: Improving biofabrication by printing stem cells for skeletal regenerative medicine. Biomaterials. 2019;**209**:10-24

[22] Kačarević ŽP, Rider PM, Alkildani S, Retnasingh S, Smeets R, Jung O, et al. An introduction to 3D bioprinting: Possibilities, challenges and future aspects. Materials. 2018;**11**(11):2199

[23] You F, Eames BF, Chen X. Application of extrusion-based hydrogel bioprinting for cartilage tissue engineering. International Journal of Molecular Sciences. 2017;**18**(7):1597

[24] Murphy R, Walsh DP, Hamilton CA, Cryan SA, In Het Panhuis M, Heise A. Degradable 3D-printed hydrogels based on star-shaped copolypeptides. Biomacromolecules. 2018;**19**(7):2691-2699

[25] Lv CF, Zhu LY, Shi JP, Li ZA, Tang WL, Liu TT, et al. The fabrication of tissue engineering scaffolds by inkjet printing technology. Materials Science Forum. 2018;**934**:129-133

[26] Jammalamadaka U, Tappa K. Recent advances in biomaterials for 3D printing and tissue engineering. Journal of Functional Biomaterials. 2018;**9**(1):22

[27] Hoque ME, Chuan YL, Pashby I. Extrusion based rapid prototyping technique: An advanced platform for tissue engineering scaffold fabrication. Biopolymers. 2012;**97**(2):83-93

[28] Bishop ES, Mostafa S, Pakvasa M, Luu HH, Lee MJ, Wolf JM, et al. 3-D bioprinting technologies in tissue engineering and regenerative medicine: Current and future trends. Genes and Diseases. 2017;**4**(4):185-195

[29] Skardal A, Atala A. Biomaterials for integration with 3-D bioprinting. Annals of Biomedical Engineering. 2015;**43**(3):730-746

[30] Cui H, Nowicki M, Fisher JP, Zhang LG. 3D bioprinting for organ regeneration. Advanced Healthcare Materials. 2017;**6**(1):1601118

[31] Ahangar P, Cooke ME, Weber MH, Rosenzweig DH, Ahangar P, Cooke ME, et al. Current biomedical applications of 3D printing and additive

manufacturing. Applied Sciences. 2019;**9**(8):1713

[32] Tappa K, Jammalamadaka U. Novel biomaterials used in medical 3D printing techniques. Journal of Functional Biomaterials. 2018;**9**(1):17

[33] Ahlfeld T, Cidonio G, Kilian D, Duin S, Akkineni AR, Dawson JI, et al. Development of a clay based bioink for 3D cell printing for skeletal application. Biofabrication. 2017;**9**(3):034103

[34] Zhang YS, Yue K, Aleman J, Mollazadeh-Moghaddam K, Bakht SM, Yang J, et al. 3D bioprinting for tissue and organ fabrication. Annals of Biomedical Engineering. 2017;**45**(1):148-163

[35] Moroni L, Burdick JA, Highley C, Lee SJ, Morimoto Y, Takeuchi S, et al. Biofabrication strategies for 3D in vitro models and regenerative medicine. Nature Reviews Materials. 2018;**3**(5):21-37

[36] You S, Li J, Zhu W, Yu C, Mei D, Chen S. Nanoscale 3D printing of hydrogels for cellular tissue engineering. Journal of Materials Chemistry B. 2018;**6**(15):2187-2197

[37] Donderwinkel I, van Hest JCM, Cameron NR. Bio-inks for 3D bioprinting: Recent advances and future prospects. Polymer Chemistry. 2017;**8**(31):4451-4471

[38] Maiullari F, Costantini M, Milan M, Pace V, Chirivì M, Maiullari S, et al. A multi-cellular 3D bioprinting approach for vascularized heart tissue engineering based on HUVECs and iPSC-derived cardiomyocytes. Scientific Reports. 2018;**8**(1):13532

[39] Park JY, Shim J-H, Choi S-A, Jang J, Kim M, Lee SH, et al. 3D printing technology to control BMP-2 and VEGF delivery spatially and temporally to promote large-volume bone regeneration. Journal of Materials Chemistry B. 2015;**3**(27):5415-5425

[40] Oh K-J, Yu HS, Park J, Lee H-S, Park SA, Park K. Co-culture of smooth muscle cells and endothelial cells on three-dimensional bioprinted polycaprolactone scaffolds for cavernosal tissue engineering. The Aging Male. 2019;**1-6**

[41] Popov VV, Muller-Kamskii G, Kovalevsky A, Dzhenzhera G, Strokin E, Kolomiets A, et al. Design and 3D-printing of titanium bone implants: Brief review of approach and clinical cases. Biomedical Engineering Letters. 2018;**8**(4):337-344

[42] Bose S, Vahabzadeh S, Bandyopadhyay A. Bone tissue engineering using 3D printing. Materials Today. 2013;**16**(12):496-504

[43] Sachs E, Cima M, Cornie J. Three-dimensional printing: Rapid tooling and prototypes directly from a CAD model. CIRP Annals. 1990;**39**(1):201-204

[44] Goriainov V, McEwan JK, Oreffo ROC, Dunlop DG. Application of 3D-printed patient-specific skeletal implants augmented with autologous skeletal stem cells. Regenerative Medicine. 2018;**13**(3):283-294

[45] Shahrokhi S, Arno A, Jeschke MG. The use of dermal substitutes in burn surgery: Acute phase. Wound Repair and Regeneration. 2014;**22**(1):14-22

[46] Varkey M, Visscher DO, van Zuijlen PPM, Atala A, Yoo JJ. Skin bioprinting: The future of burn wound reconstruction? Burns & Trauma. 2019;**7**(1):4

[47] Tarassoli SP, Jessop ZM, Al-Sabah A, Gao N, Whitaker S, Doak S, et al. Skin tissue engineering using 3D bioprinting: An evolving research field. Journal of Plastic, Reconstructive & Aesthetic Surgery. 2018;**71**(5):615-623

[48] Albanna M, Binder KW, Murphy SV, Kim J, Qasem SA, Zhao W, et al. In situ

bioprinting of autologous skin cells accelerates wound healing of extensive excisional full-thickness wounds. Scientific Reports. 2019;**9**(1):1856

[49] Jain RK, Au P, Tam J, Duda DG, Fukumura D. Engineering vascularized tissue. Nature Biotechnology. 2005;**23**(7):821

[50] Jaklenec A, Stamp A, Deweerd E, Sherwin A, Langer R. Progress in the tissue engineering and stem cell industry "are we there yet?". Tissue Engineering. Part B, Reviews. 2012;**18**(3):155-166

[51] Richards D, Jia J, Yost M, Markwald R, Mei Y. 3D bioprinting for vascularized tissue fabrication. Annals of Biomedical Engineering. 2017;**45**(1):132-147

[52] Kolesky DB, Homan KA, Skylar-Scott MA, Lewis JA. Three-dimensional bioprinting of thick vascularized tissues. Proceedings of the National Academy of Sciences of the United States of America. 2016;**113**(12):3179-3184

[53] Miller JS, Stevens KR, Yang MT, Baker BM, Nguyen D-HT, Cohen DM, et al. Rapid casting of patterned vascular networks for perfusable engineered three-dimensional tissues. Nature Materials. 2012;**11**:768

[54] Mirabella T, MacArthur JW, Cheng D, Ozaki CK, Woo YJ, Yang MT, et al. 3D-printed vascular networks direct therapeutic angiogenesis in ischaemia. Nature Biomedical Engineering. 2017;**1**:83

[55] Morrison RJ, Hollister SJ, Niedner MF, Mahani MG, Park AH, Mehta DK, et al. Mitigation of tracheobronchomalacia with 3D-printed personalized medical devices in pediatric patients. Science Translational Medicine. 2015;7(285):285ra64

[56] Kim D, Lim J-Y, Shim K-W, Han JW, Yi S, Yoon DH, et al. Sacral reconstruction with a 3D-printed implant after hemisacrectomy in a patient with sacral osteosarcoma: 1-year follow-up result. Yonsei Medical Journal. 2017;**58**(2):453-457

[57] Wu A-M, Lin J-L, Kwan KYH, Wang X-Y, Zhao J. 3D-Printing techniques in spine surgery: The future prospects and current challenges. Expert Review of Medical Devices. 2018;**15**(6):399-401

[58] Mehrotra S, Moses JC, Bandyopadhyay A, Mandal BB. 3D printing/bioprinting based tailoring of in vitro tissue models: Recent advances and challenges. ACS Applied Bio Materials. 2019;**2**(4):1385-1405

[59] Nguyen D, Robbins J, Crogan-Grundy C, Gorgen V, Bangalore P, Perusse D, et al. Functional characterization of three-dimensional (3D) human liver tissues generated by an automated bioprinting platform. The FASEB Journal. 2015;**29**(1_supplement):LB424

[60] Skeldon G, Lucendo-Villarin B, Shu W. Three-dimensional bioprinting of stem-cell derived tissues for human regenerative medicine. Philosophical Transactions of the Royal Society B. 2018;**373**(1750):20170224

[61] Lee A, Feinberg AW. 3D bioprinting of cardiac muscle tissue. In: 3D Bioprinting in Regenerative Engineering. USA: CRC Press; 2018. p. 247-268

[62] Liu J, He J, Liu J, Ma X, Chen Q, Lawrence N, et al. Rapid 3D bioprinting of in vitro cardiac tissue models using human embryonic stem cell-derived cardiomyocytes. Bioprinting. 2019;**13**:e00040

[63] Turunen S, Kaisto S, Skovorodkin I, Mironov V, Kalpio T, Vainio S, et al. 3D bioprinting of the kidney—Hype or hope? AIMS Cell and Tissue Engineering. 2018;**2**(3):119-162

[64] Subramanian B, Rudym D, Cannizzaro C, Perrone R, Zhou J, Kaplan DL. Tissue-engineered three-dimensional in vitro models for normal and diseased kidney. Tissue Engineering. Part A. 2010;**16**(9):2821-2831

[65] Kim J, Shim IK, Hwang DG, Lee YN, Kim M, Kim H, et al. 3D cell printing of islet-laden pancreatic tissue-derived extracellular matrix bioink constructs for enhancing pancreatic functions. Journal of Materials Chemistry B. 2019;**7**(10):1773-1781

[66] Lee SJ, Lee JB, Park Y-W, Lee DY. 3D bioprinting for artificial pancreas organ. Advances in Experimental Medicine and Biology. 2018;**1064**:355-374

[67] Jung CS, Kim BK, Lee J, Min B-H, Park S-H. Development of printable natural cartilage matrix bioink for 3D printing of irregular tissue shape. Tissue Engineering and Regenerative Medicine. 2018;**15**(2):155-162

[68] Lipskas J, Deep K, Yao W. Robotic-assisted 3D bio-printing for repairing bone and cartilage defects through a minimally invasive approach. Scientific Reports. 2019;**9**(1):3746

[69] Zhu W, Tringale KR, Woller SA, You S, Johnson S, Shen H, et al. Rapid continuous 3D printing of customizable peripheral nerve guidance conduits. Materials Today. 2018;**21**(9):951-959

[70] Petcu EB, Midha R, McColl E, Popa-Wagner A, Chirila TV, Dalton PD. 3D printing strategies for peripheral nerve regeneration. Biofabrication. 2018;**10**(3):32001

[71] Vogenberg FR, Isaacson Barash C, Pursel M. Personalized medicine: Part 1: Evolution and development into theranostics. Physical Therapy. 2010;**35**(10):560-576

[72] Duffy DJ. Problems, challenges and promises: Perspectives on precision medicine. Briefings in Bioinformatics. 2015;**17**(3):494-504

[73] Ginsburg GS, Phillips KA. Precision medicine: From science to value. Health Affairs. 2018;**37**(5):694-701

[74] Li X, Liu L, Zhang X, Xu T. Research and development of 3D printed vasculature constructs. Biofabrication. 2018;**10**(3):32002

[75] Di Bella C, Duchi S, O'Connell CD, Blanchard R, Augustine C, Yue Z, et al. In situ handheld three-dimensional bioprinting for cartilage regeneration. Journal of Tissue Engineering and Regenerative Medicine. 2018;**12**(3):611-621

[76] Keriquel V, Oliveira H, Rémy M, Ziane S, Delmond S, Rousseau B, et al. In situ printing of mesenchymal stromal cells, by laser-assisted bioprinting, for in vivo bone regeneration applications. Scientific Reports. 2017;**7**(1):1778

[77] Cohen DL, Lipton JI, Bonassar LJ, Lipson H. Additive manufacturing forin siturepair of osteochondral defects. Biofabrication. 2010;**2**(3):35004

[78] Witowski JS, Pędziwiatr M, Major P, Budzyński A. Cost-effective, personalized, 3D-printed liver model for preoperative planning before laparoscopic liver hemihepatectomy for colorectal cancer metastases. International Journal of Computer Assisted Radiology and Surgery. 2017;**12**(12):2047-2054

[79] Lau IWW, Liu D, Xu L, Fan Z, Sun Z. Clinical value of patient-specific three-dimensional printing of congenital heart disease: Quantitative and qualitative assessments. PLoS One. 2018;**13**(3):e0194333

[80] Schöneberg J, De Lorenzi F, Theek B, Blaeser A, Rommel D, Kuehne AJC, et al. Engineering biofunctional in vitro vessel models using a multilayer

bioprinting technique. Scientific Reports. 2018;**8**(1):10430

[81] Kim BS, Lee J-S, Gao G, Cho D-W. Direct 3D cell-printing of human skin with functional transwell system. Biofabrication. 2017;**9**(2):25034

[82] Berry DB, You S, Warner J, Frank LR, Chen S, Ward SR. A 3D tissue-printing approach for validation of diffusion tensor imaging in skeletal muscle. Tissue Engineering. Part A. 2017;**23**(17-18):980-988

[83] Bourland J, Fradette J, Auger FA. Tissue-engineered 3D melanoma model with blood and lymphatic capillaries for drug development. Scientific Reports. 2018;**8**(1):13191

[84] Albritton JL, Miller JS. 3D bioprinting: Improving in vitro models of metastasis with heterogeneous tumor microenvironments. Disease Models & Mechanisms. 2017;**10**(1):3-14

[85] Zhang YS, Duchamp M, Oklu R, Ellisen LW, Langer R, Khademhosseini A. Bioprinting the cancer microenvironment. ACS Biomaterials Science & Engineering. 2016;**2**(10):1710-1721

[86] Dai X, Ma C, Lan Q, Xu T. 3D bioprinted glioma stem cells for brain tumor model and applications of drug susceptibility. Biofabrication. 2016;**8**(4):045005

[87] King SM, Presnell SC, Nguyen DG. Abstract 2034: Development of 3D bioprinted human breast cancer for in vitro drug screening. Cancer Research. 2014;**74**(Suppl 19):2034-2034

[88] Bakhtiar SM, Butt HA, Zeb S, Quddusi DM, Gul S, Dilshad E. 3D printing technologies and their applications in biomedical science. In: Omics Technologies and Bio-Engineering: Towards Improving Quality of Life. UK: Elsevier; 2017

[89] Min D, Lee W, Bae IH, Lee TR, Croce P, Yoo SS. Bioprinting of biomimetic skin containing melanocytes. Experimental Dermatology. 2018;**27**(5):453-459

[90] Lee V, Singh G, Trasatti JP, Bjornsson C, Xu X, Tran TN, et al. Design and fabrication of human skin by three-dimensional bioprinting. Tissue Engineering. Part C, Methods. 2014;**20**(6):473-484

[91] Tseng H, Gage JA, Shen T, Haisler WL, Neeley SK, Shiao S, et al. A spheroid toxicity assay using magnetic 3D bioprinting and real-time mobile device-based imaging. Scientific Reports. 2015;**5**:13987-13987

[92] Isaacson A, Swioklo S, Connon CJ. 3D bioprinting of a corneal stroma equivalent. Experimental Eye Research. 2018;**173**:188-193

[93] Ghezzi CE, Marelli B, Omenetto FG, Funderburgh JL, Kaplan DL. 3D functional corneal stromal tissue equivalent based on corneal stromal stem cells and multi-layered silk film architecture. PLoS One. 2017;**12**(1):e0169504-e0169504

[94] Madden LR, Nguyen TV, Garcia-Mojica S, Shah V, Le AV, Peier A, et al. Bioprinted 3D primary human intestinal tissues model aspects of native physiology and ADME/Tox functions. iScience. 2018;**2**:156-167

[95] Mozetic P, Giannitelli SM, Gori M, Trombetta M, Rainer A. Engineering muscle cell alignment through 3D bioprinting. Journal of Biomedical Materials Research Part A. 2017;**105**(9):2582-2588

[96] Chen S, Nakamoto T, Kawazoe N, Chen G. Engineering multi-layered skeletal muscle tissue by using 3D microgrooved collagen scaffolds. Biomaterials. 2015;**73**:23-31

[97] Zhang YS, Arneri A, Bersini S, Shin SR, Zhu K, Goli-Malekabadi Z, et al. Bioprinting 3D microfibrous scaffolds for engineering endothelialized myocardium and heart-on-a-chip. Biomaterials. 2016;**110**:45-59

[98] Lind JU, Busbee TA, Valentine AD, Pasqualini FS, Yuan H, Yadid M, et al. Instrumented cardiac microphysiological devices via multimaterial three-dimensional printing. Nature Materials. 2017;**16**(3):303-308

[99] Guguen-Guillouzo C, Corlu A, Guillouzo A. Stem cell-derived hepatocytes and their use in toxicology. Toxicology. 2010;**270**(1):3-9

[100] Hewitt NJ, Lechón MJG, Houston JB, Hallifax D, Brown HS, Maurel P, et al. Primary hepatocytes: Current understanding of the regulation of metabolic enzymes and transporter proteins, and pharmaceutical practice for the use of hepatocytes in metabolism, enzyme induction, transporter, clearance, and hepatotoxicity studies. Drug Metabolism Reviews. 2007;**39**(1):159-234

[101] Khetani SR, Bhatia SN. Microscale culture of human liver cells for drug development. Nature Biotechnology. 2008;**26**(1):120-126

[102] Kang K, Kim Y, Jeon H, Lee SB, Kim JS, Park SA, et al. Three-dimensional bioprinting of hepatic structures with directly converted hepatocyte-like cells. Tissue Engineering. Part A. 2018;**24**(7-8):576-583

[103] Ma X, Qu X, Zhu W, Li Y-S, Yuan S, Zhang H, et al. Deterministically patterned biomimetic human iPSC-derived hepatic model via rapid 3D bioprinting. Proceedings of the National Academy of Sciences of the United States of America. 2016;**113**(8):2206-2211

[104] Mironov V, Boland T, Trusk T, Forgacs G, Markwald RR. Organ printing: Computer-aided jet-based 3D tissue engineering. Trends in Biotechnology. 2003;**21**(4):157-161

[105] Ventola CL. Medical applications for 3D printing: Current and projected uses. Physical Therapy. 2014;**39**(10):704-711

[106] Yin L, Bien H, Entcheva E. Scaffold topography alters intracellular calcium dynamics in cultured cardiomyocyte networks. American Journal of Physiology-Heart and Circulatory Physiology. 2004;**287**(3):H1276-85

[107] Rai J, Kaushik K. Reduction of animal sacrifice in biomedical science & research through alternative design of animal experiments. Saudi Pharmaceutical Journal. 2018;**26**(6):896-902

Section 4

Agricultural Applications

Chapter 12

Application the Geometric Modeling Methods and Systems in Design Engineering and Manufacturing on Example of Agriculture Engineering

Tojiddin Juraev Khayrullaevich,
Murodov Nusrat Murtazoyevich
and Naimov Sandjar Tulkunovich

Abstract

Geometric modeling, as widely used synthetic design method in design engineering and manufacturing, is a theoretical base for geometric modeling application (industrial design) and geometric modeling systems (CAD systems). Therefore this chapter is devoted to application geometric modeling methods and systems in design engineering and manufacturing. For example, we will considered designing agriculture machines' tools by following case studies: development models of bulldozer's moldboard by geometric modeling method (for design engineering); screening the concept select process of plow's moldboard (for design engineering and manufacturing); determining integrative role of geometric modeling systems in agro machinery tools' PLM (for manufacturing).

Keywords: geometric modeling methods and systems, geometric modeling application, product lifecycle, design engineering, multifunctionality, moldboard's surface, product manufacturing criterions and properties

1. Introduction

Today, world has seen a marked increase in the variety of techniques used in all areas of human activity, in terms of their functionality. Therefore working out of working bodies expanding their functionality is one of the major problems of modern engineering and design activity. This has become especially important in the context of the transition of the Republic of Uzbekistan to a new stage of development, where fundamental reforms of management in agriculture are taking place, grandiose projects in construction are being implemented, as well as great attention is being paid to the development of settlements [1]. Decision of these problems is straight connected with geometric modeling on which modern problems of design engineering and manufacturing are based. Extending the functional possibilities of agricultural machines is one of the main ways for design engineering and manufacturing of modern agricultural machines, which can be realize by design engineers. Particularly, it actually in modern agricultural production, which based on precision agriculture technologies, where using such machines must provide: reduction of cost, conservation of ground fertility, saving energy-resources, improvement labor conditions, and increase machines capacity. One of the efficient ways to decide these problems is using geometric modeling in design engineering of agriculture machines' tools. Geometric modeling, as one of varieties synthetic methods' of designing, is a theoretical base for production design and CAD technologies, which widely use in design engineering and manufacturing [2–4].

The development of agricultural tools mankind has been engaged for centuries. In 1830, the Italian abbes Lambruschini and Ridolfi were prompted by the helical surface (helix) to plow's moldboard. In the late nineteenth century, Russian academician V.P. Goryachkin laid the foundation of the science "Agricultural Mechanics". Since then, many scientists from America, Europe, and other countries have developed various methods of research and modeling in the mechanization of agriculture. These methods mainly solve problems by analytical and experimental methods, although synthetic methods have a number of advantages in solving some problems. The reason that synthetic methods, as geometric modeling, until recently used rarely, was the advantage of the use of information technology in analytical methods. However, the widespread use of information technology in synthetic methods since the end of the twentieth century, their capabilities have become much more effective. The advantage of geometric modeling is its simplicity and clarity. It as a basis of synthetic methods of development of technical objects has innovative character that modern production demands increase in a variety of production, reduction of terms of their development, and also automatization of these processes [5–8].

2. Developing the Bulldozer's moldboard by geometric modeling method for design engineering

2.1 Designing directions of mold board's working surface

Considering decision of above-mentioned problem on example of moldboard type tools give clarity on this problem. Moldboards, as main working body of bulldozers, graders, and other special equipment, are designed to perform preparatory work in agriculture and melioration, ground works in road building and engineering preparation of territories, as well as other works, for example, in municipal service. Classic moldboard has frontally positioned cylindrical working surface, on which the earth or other mass must move, formed as a "dragging prism" in the required direction and quantity [9, 10]. To expand functionality of moldboards, there are have developments in various design options, with a changing position of the working surface or using other working surface (**Table 1**). But, these developments' directions aimed to expand functionality of moldboards, to perform definite works [9, 11]. The decision of these problems is straight connected with geometric modeling which is based on modern problems of industrial design [2–4]. The result of using industrial design at development of moldboard type tools on base of constructive geometric modeling is a "design-project" of moldboard, which possibly produce three working surface types design. Design-project is geometric model of developing object which has only geometric parameters, but these parameters are given on base of forward given conditions of technical/technological parameters and have close connections with them. We shall consider the design-project of

DOI: http://dx.doi.org/10.5772/intechopen.89974

No.	Moldboards' construction	Surface type	No.	Moldboards' construction	Surface type
1		Cylindrical (Poland)	4		Combined (Czech)
2		Conidial (Finland)	5		Cylindroidal (Sweden)
3		Conical (USA)	6		Planar (France)

Table 1.
Types of bulldozer moldboards' construction and their surface.

moldboard's working surface consisting of section. For base of the models, we take multi-function surface consisting linear surfaces, which are broadly used for designing moldboards (**Table 2**). Design-project of mold board's working surface by constructive geometric model applicable for work execution of characteristics: technical, technological, and economical factors of designed technology, allows more flexible control its functional possibility, solving constructive problems [2, 12, 13]. The analysis of existing moldboard design and studies upon their improvement shows that creation new design increasing their functional possibilities, way of constructive geometric modeling, have a broad prospect [9, 11, 12, 14, 15]. With standpoint of the constructive geometry design of moldboard's working surfaces will possible divide into three types: (1) construction consisting traditional (one-piece) surface design (**Figure 1**); (2) construction consisting sectional (parts) surface design (**Figure 2**); and (3) construction consisting elemental (plates) surface design (**Figure 3**). Here-with possible sweeps away prospects of primary using these design on example: (1) traditional surface design for producing polymeric moldboards; (2) sectional surface design for expansion of functional possibilities and increasing manufacturability of moldboards' producing; and (3) elemental surface design for best managing manufacture, functional, working, and other quality moldboards. Development of moldboard's working surface, applicable to execution of different works, increases their operational,

No.	Geometry of moldboard's surface	Using in machines
1	Frontal planar surface	Channel defogger's moldboard
2	Inclined planar surface	Bush cutting bulldozer's moldboard
3	Frontal cylindrical surface	Frontal bulldozer's moldboard
4	Inclined cylindrical surface	Bucket scraper's moldboard
5	Frontal conical surface	Grader's moldboard
6	Inclined conical surface	Frontal plow's moldboard
7	Cylindroidal surface	Universal plow's moldboard
8	Conidial surface	High-speed plow's moldboard
9	Hyperbolic-parabolic surface	Hyperbolic body plow's moldboard
10	Helicoid surface	Helicoids body plow's moldboard
11	Torso surface	Cultural plow's moldboard
12	Combined surface	Combined body plow's moldboard

Table 2.
Geometry of moldboard's surface and their using in machines.

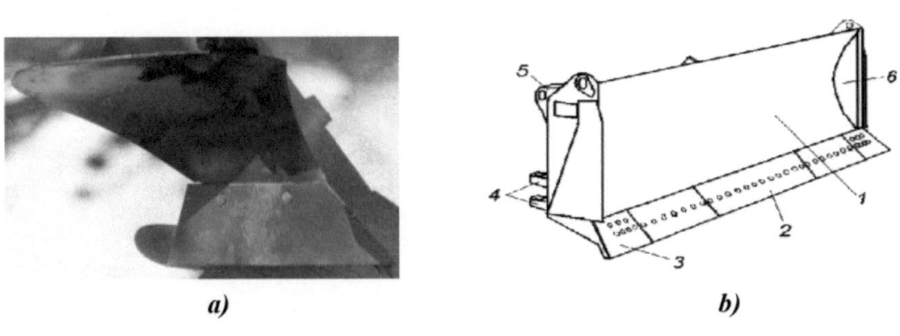

Figure 1.
Traditional (one piece) construction of plow (a) and bulldozer (b) moldboard's surface.

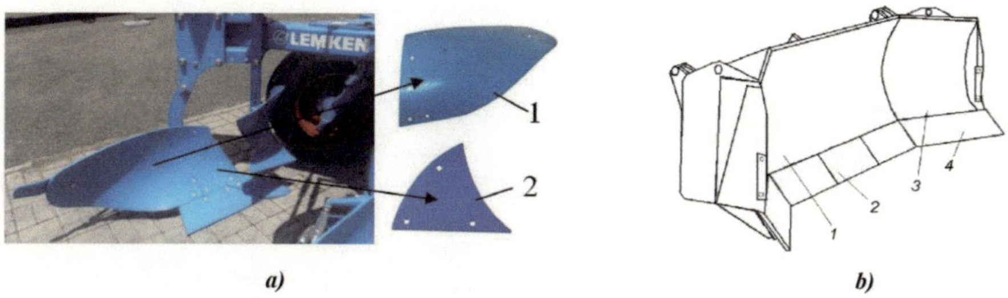

Figure 2.
Sectional (parts) construction of plow (a) and bulldozer (b) moldboard's surface: "1-wing" and "2-breast" of plow's body; "2-frontal" and "3-side" sections of spherical moldboard of bulldozer.

Figure 3.
Elemental (plates) construction of plow (a) and bulldozer (b) moldboard's surface: 1-right, 2-middle and 3-left guiding frames, 4-right and 5-left formative plates (*construction offered by author).*

economic, and technological performance. Therefore, the design-project of a constructive geometric model of the moldboard's working surfaces, although there are designs of such equipment, allows more flexibility manage the functional capabilities of moldboard and solve above-described design problem [2, 12, 13].

2.2 Geometric modeling of transformable moldboard's surface

Linear surfaces are main use type of in moldboards' working surface. Lines l are formatives of cylindrical working surface Φ, and all of them are parallel to each other. In considering task they have horizontally position. Working surface Φ is formed by directory curve m. Type of this curve is planar and it can be given by plane P. For frontal moldboard, the plane P is located perpendicular to formative lines in the middle of them. This plane intersects the working surface Φ, and divides it into two equal Φ_a and Φ_b parts, simultaneously being the symmetry plane of these working surfaces. Let us choose the straight line k on the symmetry plane, through which we can carry out the beam of planes. These planes intersect with the Φ_a and Φ_b working surfaces to form intersection curves. Let us define these planes on both sides of the symmetry plane P, respectively, $P_1, P_2, \dots, P_n$ and $P_1', P_2', \dots, P_n'$, as well as the intersection lines on the working surfaces Φ_a and Φ_b, respectively, curves m_1, $m_2, \dots, m_n$ and $m_1', m_2', \dots, m_n'$. In this case, the angles between planes and plane of symmetry P, respectively, denote $\alpha_1, \alpha_2, \dots \alpha_n$. Each pair of curves m_1, m_1'; m_2, m_2'; ... m_n, m_n', formed, respectively, by pairs of planes $P_1, P_1'; P_2, P_2'; \dots P_n, P_n'$, are symmetrical, where k is the axis of mirror reflection of pairs of curves on working surfaces Φ_a and Φ_b (**Figure 4a**). Therefore, when the pairs of P_i and P_i' planes rotate together with Φ_a and Φ_b surfaces around the k-axis by the corresponding angle α_i, P_i, and P_i' planes, as well as their m_i and m_i' curves are match and form a

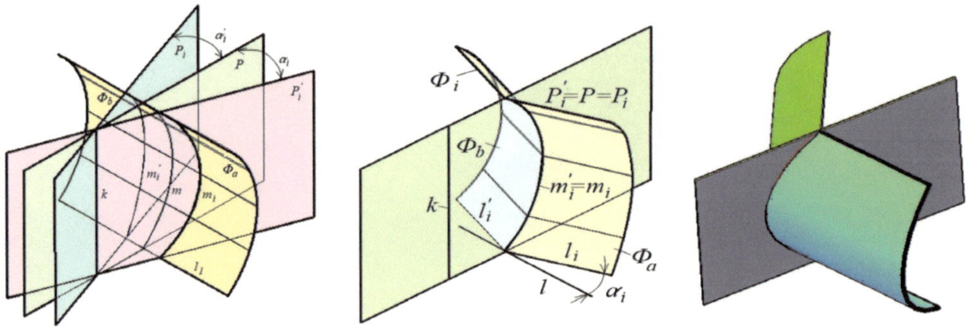

a) *b)* *c)*

Figure 4.
Developing the conception of moldboard with bilateral action working surface. a) geometric model for designing moldboard with bilateral working surface; b) moldboard for outside action; c) moldboard for inside action.

new working surface. As a result, an edge separating the working surface into two halves is formed on the working surface. On the basis of this model, it is possible to develop various constructive variants of transformed moldboard's design models, allowing to transform from one working surface in another. It is known that when designing complexity technical forms, the considering surface is mentally different by "*geometric*" and "*working*," since from the same surface, it is possible to obtain different working surfaces [4, 11]. Therefore, it can form new required working surface $\boldsymbol{\Phi_i}$ by proposed model, that is, by rotating working surfaces $\boldsymbol{\Phi_a}$ and $\boldsymbol{\Phi_b}$ around the axis $\boldsymbol{k}$ in angle $\boldsymbol{\alpha}_i$. Although given $\boldsymbol{\Phi}$ and newly formed $\boldsymbol{\Phi_i}$ surfaces are cylindrical, they have different working surfaces with different functional properties, where $\boldsymbol{\alpha}$ becomes the control parameter in the formation of $\boldsymbol{\Phi_i}$. New working surface $\boldsymbol{\Phi_i}$ improves directional action of the moving layer mass on the outside (**Figure 4b**) and from the inside (**Figure 4c**) than given surface $\boldsymbol{\Phi}$.

2.3 Geometric parameterization of moldboard's surface

2.3.1 Giving the axis of rotation of the working surface

The process of formation of the required working surface $\boldsymbol{\Phi_i}$ can be controlled, besides the parameter $\boldsymbol{\alpha}$, also the position $\boldsymbol{k}$. In the model under consideration, position of rotation axis $\boldsymbol{k}$ is vertical and has a certain distance relative to $\boldsymbol{\Phi_i}$. However, change in the position of $\boldsymbol{k}$ significantly affects the formation of $\boldsymbol{\Phi_i}$. Here we can consider two parameters of $\boldsymbol{k}$: the change in the distance $\boldsymbol{f}$ defined between fixed points $\boldsymbol{k}$ and $\boldsymbol{m}$, for example, the base of $\boldsymbol{k}$ and sock $\boldsymbol{m}$ on a horizontal plane; and the change in the angle of inclination $\boldsymbol{\beta}$ to the horizontal plane. At the same angle $\boldsymbol{\alpha}_i$ and shape of the directory curve $\boldsymbol{m_i}$, changing $\boldsymbol{f}$ will lead to change in the relative position of pairs of directory curve $\boldsymbol{m}_i$ and $\boldsymbol{m}_i'$, which will lead to a change in the design parameters of moldboard's working surface $\boldsymbol{\Phi_i}$. Among the options (**Figure** 5) considered by the author, the variants b) chord $\boldsymbol{AB}$ and d) tangent in point **C** are selected as acceptable for this problem, when it will be possible to neglect parameter $\boldsymbol{f}$, that simplifies the problem. Though other variants also have such working surface, it they can lead to complication of moldboard's constructive parameters. However, when the surface $\boldsymbol{\Phi_i}$ is formed, in the variant *d* rotation is performed in the opposite direction than in variant *b*. The rotation angle $\boldsymbol{\alpha}$ is selected with $\boldsymbol{0 < \alpha < \alpha_{max}}$, provided that the $\boldsymbol{P}_i$ and $\boldsymbol{P}_i'$ planes must intersect all formative lines of surfaces $\boldsymbol{\Phi_i}$, where $\boldsymbol{\alpha_{max}}$ is equal to $\boldsymbol{tg\alpha = (l/2)/b}$, and $\boldsymbol{b}$–extension of directory curve.

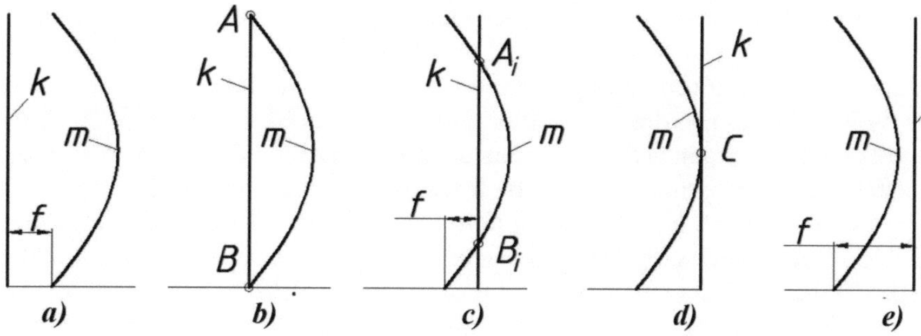

Figure 5.
*Variants of rotating axis (**k**) positions of working surface relatively to directory curve **m**. A, B, C – extremely points of working surface, **f** – distance between axis and extremely points.*

2.3.2 Parameters of designed working surface's directory curve

It is necessary to mark the parameters by shape and position of the directory curve ***m*** of the surface Φ. According to the problem, the shape of directory curve ***m*** is flat and smooth, with a certain curvature and a concave side forward. Since these properties of directory curve remain low during the transformation of the surface, they will be identified as the topological parameters of the curve that determine its shape. Therefore, such surface parameters as its shape and curvature also remain low even when a new surface Φ_i is formed. The position of the curve is defined by two parameters: its offset–***b*** and height–***h*** of the curve. They are defined as constructive parameters, as they define the design of moldboard. The following variants of mutual arrangement of constructive parameters ***m***, determined by the position of characteristic points, can be distinguished (**Figure 6**). The lower (***A***) and upper (***B***) points define ***h***, and the outermost left and right (pairs of from points ***A***, ***B***, ***C***) points define ***b***. These directory curve variants can be selected when designing the moldboard depending on the work performed by it. When ***f*** is changed in vertical position ***k***, the moldboard's overall height ***h′*** also remains low. Parameter $\delta b_{max} = b_i - b$ obtained after formation of an edge of the surface Φ_i edge, is located opposite to the point at which the rotation axis ***k*** passes (*right/left*–on the chest or *upper/lower*–on the toe).

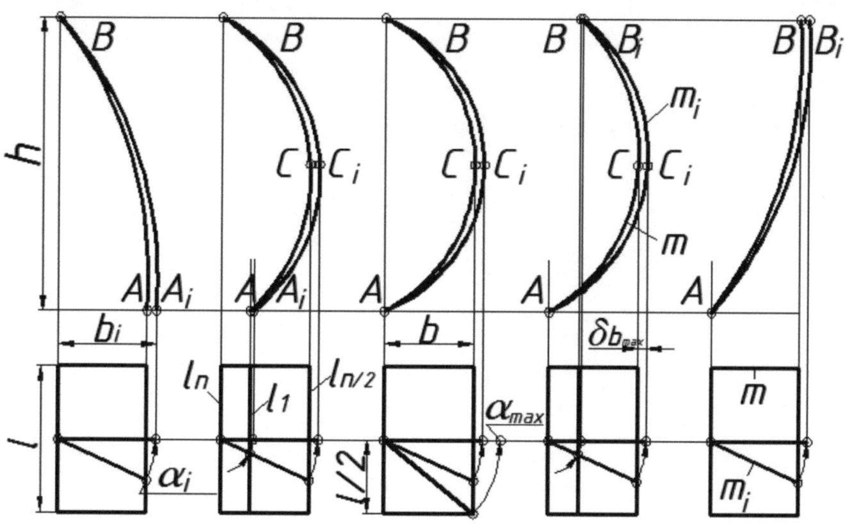

Figure 6.
Variants of relative positions of directory curve and it's constructive parameters.

2.3.3 Parameters of designed working surface's formative lines

Criteria for choosing the variants of the characteristic points of the directory curve ***m*** by ***h*** and ***b***, in the design of the moldboards can be explained by linking these points to the characteristic positions of the formative lines ***l***. For example, we distinguish the following positions of the formative lines ***l***, passing through characteristic points ***m*** in width ***b*** with respect to ***h***: upper, lower, frontal, rear, and middle (by ***h*** or ***b***), and determine their influence on the nature of the movement of the layer mass on the working surface of the moldboard (**Table 3**). It follows from **Table 3** that the nature of the layer mass movement along the working surface can be controlled by changing the relationship ***h*** and ***b***, by changing angle $\boldsymbol{\beta}$ of the ***k***-axis inclination. In contrast to vertical position, the inclination ***k*** at angle $\boldsymbol{\beta}$ forwards or backwards gives the working surface, in addition to improving the shift of the transported mass to the side when it is horizontally leveled (**Figure 7a**), also improves the functional properties of the inclined slopes (**Figure 7b**) and lifts (**Figure 7c**) from the transported mass. This is achieved by changing the positions of formative lines ***l***, which also represent the plowshares, relative to the horizontal plane by angle $\boldsymbol{\varphi}$, after formation $\boldsymbol{\Phi_i}$. The angle $\boldsymbol{\varphi}$ can be determined by the projection model based on the principles of descriptive geometry [16], superimposing horizontal plane with frontal, by rotating it in 90°, on the front projection we combine projections ***k*** and ***l*** (**Figure 7d**). Rotate ***l*** by the angle $\boldsymbol{\alpha_i}$, marking with $\boldsymbol{l'}$, and easily find the front projection $\boldsymbol{l_v'}$. Since ***l*** rotates on a frontal projection plane perpendicular to ***k***, the rotation circle ***l*** is projected on a horizontal plane as an ellipse. Using the projecting rays, we find $\boldsymbol{l_h'}$ and determine the $\boldsymbol{\varphi}$—the angle of

No.	On width *b*, in respect of *h* and through points			Nature of the moving the moveable mass on worker of the surfaces
	Anterior	**Average**	**Posterior**	
1	Superior–*B*	Not available	Interior–*A*	Powerfully postponed in before.
2	Superior–***B***	Interior–***A***	Average–***C***	Partly is taken on breast and powerfully postponed in before.
3	Superior/ Interior–***B/A***	Not available	Average–***C***	Completely taken on breast and powerfully postponed in before.
4	Interior–***A***	Superior–***B***	Average–***C***	Completely taken on breast and weakly postponed in before.
5	Interior–***A***	Not available	Superior–***B***	Completely taken on bosom.

Table 3.
Formatives' positions and their influence to working surface nature.

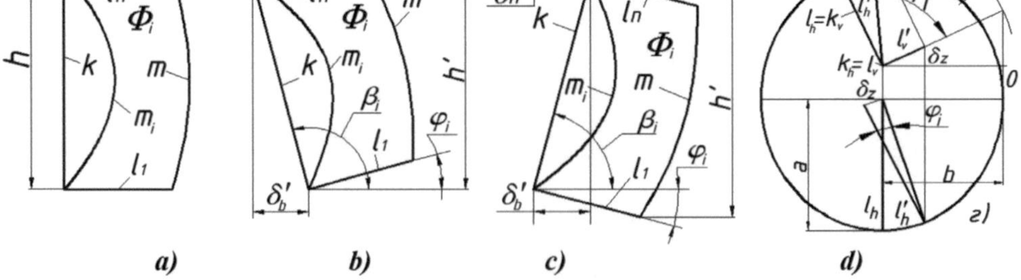

Figure 7.
Determining the geometric parameters of inclined working surface. Working surface positions: a) vertically; b) inclined to forward; c) inclined to back; d) geometric model for determining parameters.

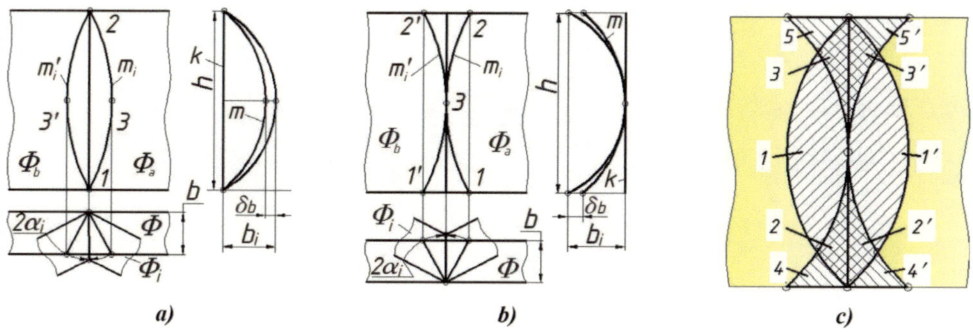

Figure 8.
Developing the construction of moldboard with bilateral action working surface. a) for outside action; b) for inside action; c) sections of working surface.

inclination l to the horizontal plane by the rectangular triangle method, also considered as the angle of inclination to the plowshare. After transformation of the working surface Φ by Φ_i with inclined k, the overall height of the blade h' increases, although h decreases by δh. The upper part of the ridge tilts forward or backward relative to the lower part, shifting by $\delta b'$. As a result of the transformation of the working surface, the lengths of corresponding l_i formatives change within the range of $0 < \delta b < \delta b_{max}$, displacing the ends of the surfaces forming the ribs, change. At the point at which the axis of rotation k passes, the length of l_i is equal to $\delta b = 0$ and the nose (upper or lower) part of it is equal to $0 < \delta b < \delta b_{max}$.

2.4 Sections of designed working surface

The definable parameters of Φ_i has two variants, on base of descriptive geometry principles, make sure that under alike α_i parameters Φ_i is also alike, but mutually negative (**Figure 8a, b**) [16]. This allow to combine two variants in one construction, which will enlarge the functional possibilities of designed moldboard (**Figure 8c**). It will select five compartments of working surfaces on intersection lines. Alternate switching-on or switching-off corresponding compartments will enable to work moldboard in three modes: moving the layer mass frontal, outside, and inside. The proposed geometric model of transformed working surface allows to development multifunctional moldboard. This development is intended for designing organization to production of specific machines. Parameterization of moldboard's working surface relieves designer's work, increases choosing variants under development moldboard's working surface, and allows effectively solve the constructive problems.

2.5 Dynamic model of working surface's directory curve

The author developed a geometric model for giving directory curve of working surface and implemented in *AutoCAD 2012* and *SIMPLEX* systems. It was found that *SIMPLEX* system has some advantage over *AutoCAD* in solving constructive geometric modeling problems [17]. Unlike *AutoCAD*, where the giving process of directory curve automated only for ellipse and circles, in *SIMPLEX* process of giving automated for any conics and Bezier curves, by the same conditions. In addition, unlike the dynamic block developed in *AutoCAD*, the dynamic model in *SIMPLEX* will not only automate the process of changing the curve parameters, but also the process of determining the projection of the guide in a different position for cylindrical surface, and in case of a cylindroidal surface will determine them as template lines in each section.

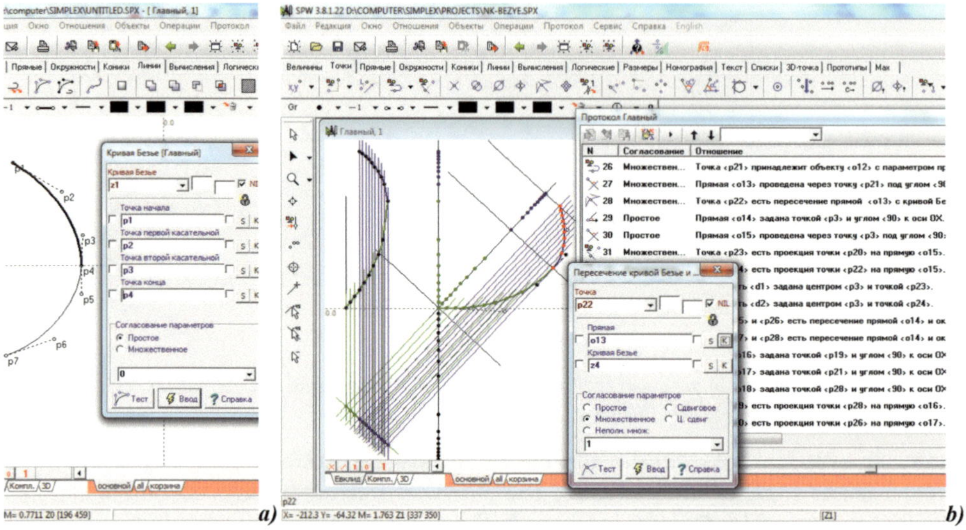

Figure 9.
Bezier curve by forward given conditions (a) and determining it's projections by multiple agreement points (b).

The plane of curve is vertical and at an angle γ_0 to the axis OX, as to the wall of the furrow. The horizontal projection of curve defines as a segment, equal to the length of L. Aligning the curve plane with horizontal plane, gives natural dimensions of curve in the plan. For easy control of curve, it can be set even by two Bezier curves. In this case, the tangent to the intermediate point P_4, is tangential simultaneously to the two Bezier curves $P_1P_2P_3P_4$ and $P_4P_5P_6P_7$ (**Figure 9a**). Determine the position of points P_1, P_4, and P_7 on the horizontal projection. Determine the frontal projections of these points by *interactive incidence* at surface formatives' heights h_0, h_i, and h_{max}. To determine the frontal projection, use the possibility of the system "*belonging to the point of the curve in multiple agreement*". In this case, this alignment can be set on a horizontal projection or on a projection in the plan. Next, the projection links determine the frontal projection (**Figure 9b**).

3. Screening the concept selecting process of Plow's moldboard model for design engineering and manufacturing

As noted in the introduction, current temps of manufacturing require development and implementation of innovative design technologies such as industrial design. The key role in this are played design engineers, because that solution of issues such energy and resource saving, increase functionality and productivity, as well as manufacturability largely depend on the technical means developed by them [3]. Studies show that use of industrial design as a powerful weapon of design engineers can give significant results in development of technical means according to various criteria, which is an applied aspect of geometric modeling [4, 18–20]. The application industrial design in this process requires the identification of tasks, the solutions of which are associated with its basis, that is, with geometric modeling [2].

The result of application industrial design in development of technical objects is their "design-project" [2, 3]. Design-project as a method of non-experimental design allows to develop technical objects with the least amount of time, labor, and money which spend in this process [18]. Therefore, the purpose of this design-project is application of research results on geometric modeling of plow's moldboard and its surface, for their further adaptation into manufacturing. Therefore, one of final results of research is the development of design-project of plug's moldboard, prepared on the basis of developed geometric models, algorithms, and methods by non-traditional design. It can highlight the following stages of design-project:

1. Analysis, evaluation, and model selection for design-project;

2. Development of the concepts and models for design-project;

3. Preparation description and sketch of proposed design;

4. Geometric modeling of the proposed design;

5. Computer simulation of the proposed design;

6. Patenting design-project for manufacturing by design conditions; and

7. Adaptation design-project for manufacturing by design conditions.

As an example, let us take a look at the design-project of plow's moldboard. At it is known, plows used in agricultural production have different design models of their moldboards according to their purpose. Each design models has its own advantage, the isomorphic application of which in another design can lead to some loss of perfection of this design. In such cases, it is possible to combine the advantages of considered models, according to various evaluation criteria, into one new design, with the necessary changes, on one of basis methods of industrial design–"*Concept selection*" [2, 20].

Moldboards have a complicated technical form, centuries-old changes to improve their designs and have a universal geometric model. These factors allow the application industrial design in the development of moldboard by geometric characteristics that affect to their technical/technological characteristics. Let us consider directions of plow's moldboard improvement.

Classic plows, having a one side turnover moldboard with cylindroidal working surface, have common use in agricultural production. Therefore, they will be considered as basic model, as it is chosen by experts as basis for development of other moldboards' design. They have good crumbling and turning indices, but these advantages are opposed to their shortcomings, which led to three improvement direction. Firstly, low manufacturability of such cases with non-sweep working surface led to development geometrically combined working surfaces [10]. Secondly, use of one side turning moldboard will lead to formation of furrows and ridges, that is, roughness of plow, which led to development vertical reversible plow. They differ in higher productivity and quality of performed works which are not demanding additional presuming agro technical actions after their using. But presence of double (right and left turning) moldboard makes construction more expensive, more metal quantity, and with greater traction resistance, which is its drawbacks, in contrast to its advantages [10]. This led to improvement in the third direction, that is, to development horizontal turn plows. There has development of technological scheme of plow with opportunities working in two right and left side, but with cylindrical working surface, which does not provide a satisfactory layer turnover [14, 21]. This analysis shows that the main reason for improvement and crossing point of advantages and disadvantages of considered designs is geometry of moldboard's working surface.

Based on of existing plows' design models and their research on improvement show that opportunity of creating design model consisting from combination of two or more design models is not used enough, and such design model combines their advantages and eliminates their disadvantages gives a solution to this problem. Among many works devoted to this problem, as an example, we can consider works relating to plow's design model [14], to technical complexity surfaces [15] or improvement parts of moldboard [11, 12]. Development was conduct on the main types of plow's moldboard design models, which takes into account several basic criteria for choice design model by geometric characteristics of moldboards, according to requirements of manufacturers and consumers [4, 9–12, 14] (**Table 4**). Among the evaluation criterions of plow, depending on its geometric parameters, it is possible to note metal quantity of construction, manufacturability of moldboard, as well as functional working quality [10]. But to combine all these quality characteristics together is problematic, because design model of moldboard has complexity geometric parameters. Production design is application of geometric modeling, and this problem can be resolve by geometric modeling.

Evaluation of criterions produced on relative to basic design model "***A***" (**Table 5**). Geometric characteristics are evaluated by their advantages (+) and disadvantages (−). The characteristics of design models that are clearly not distinguished by experts as advantages or disadvantages are conditionally evaluated neutrally (0), for reasons that they do not particularly affect to choice the design model.

Basic design model "A"		**Revised design model "D"**	
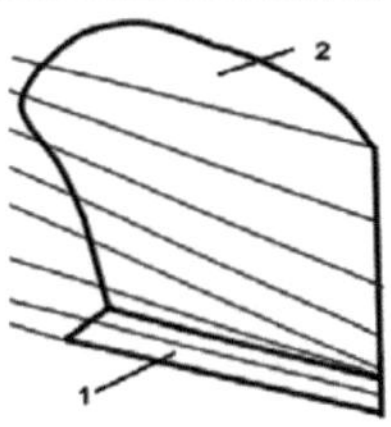	*For one side turning plow.* Moldboard type—classic body. Working surface—cylindroidal. *Plowshare (1), Moldboard (2).*	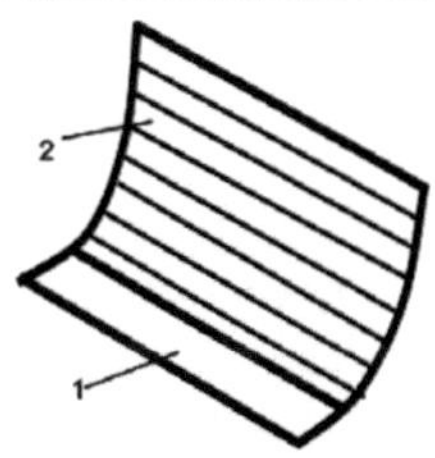	*For horizontal turn plow.* Moldboard type—double action body. Working surface–cylindrical. *Plowshare (1), Moldboard (2).*
Revised design model "B"		**Revised design model "E"**	
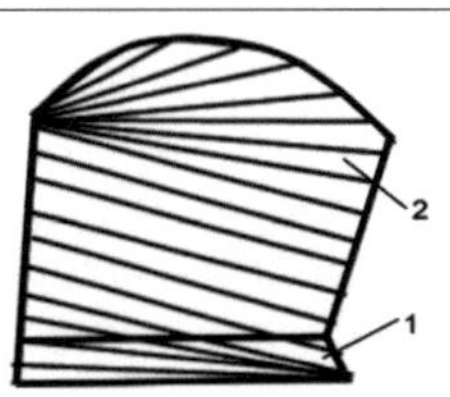	*Plow for one side turning.* Moldboard type—experimental body. Working surface—geometrical combined. *Plowshare (1), Moldboard (2).*	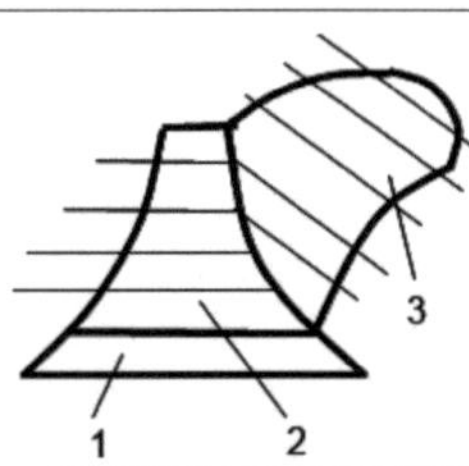	*Plow for one side turning.* Moldboard type—partial body. Working surface—geometrical combined. *Plowshare (1), breast (2), wing(3).*
Revised design model "C"		**Revised design model "F"**	
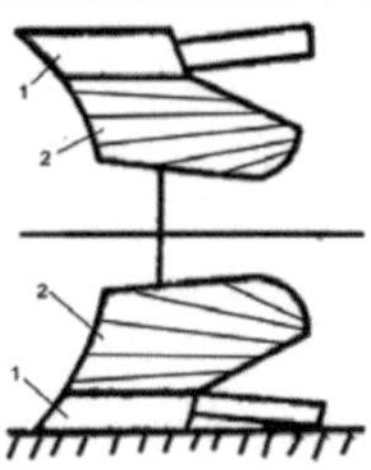	*For vertical revolving plow.* Moldboard type—doubled right and left turn body. Working surface—cylindroidal. *Plowshare (1), Moldboard (2).*	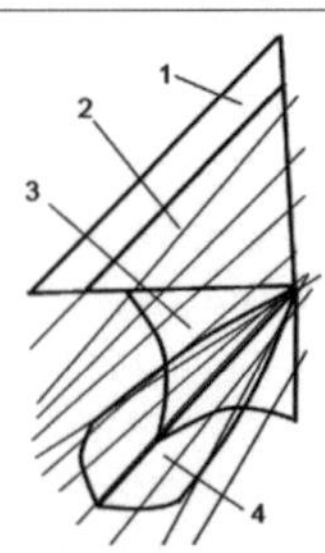	*For horizontal turn plow.* Moldboard type—partial body. Working surface—geometrical combined. *Plowshare (1), breast (2), right (3) and left (4) wings.*

Table 4.
Visual analyzing the designs of reviewed moldboard's models.

No.	Influence geometry of moldboards to technical and technological characteristics	Design models of moldboards						
		A	B	C	D	E	F	G1–G3
1	Influence to trajectory of layer	0	+	0	–	0	0	**B**
2	Influence to material quantity of plow	+	+	–	0	+	–	**BE**
3	Influence to operation quality of plowing	0	+	0	–	0	0	**B**
4	Influence to smoothness of plow	–	–	+	+	–	+	**CDF**
5	Influence to functionality of plow	–	–	–	–	+	+	**EF**
6	Influence to manufacturability of moldboard	–	–	–	+	0	0	**D**
7	Influence to complexity of moldboard design	+	+	0	0	0	–	**B**
8	Amount of disadvantages "–"	Basic	3	3	3	1	2	Proposed
9	Amount of advantages "+"		4	1	2	2	2	
10	Summarized amount		1	−2	1–	1	0	
11	Rating placement		1	4	3	1	2	

Table 5.
Evaluation the moldboards' criterions by geometric characteristics.

Material quantity and complexity of design model "*A*" has advantages, but it does not provide smooth plowing and its cylindroidal surface is low technologically for manufacturing. The functionality of this design model is also low, as it is developed for individual agricultural conditions. Therefore, its shortcomings have led to the development of new designs aimed at their elimination. Geometrically combined working surface design model "B" of experimental moldboard allows good control of layer trajectory and quality of processing. Other qualities of design model "*B*" are approximately the same with design model "*A*." They are not widely used, because their working surface has geometric complexity for giving by manufacturers. Design model "*C*" of vertical revolving plow's moldboard is the same as design model "*A*.". Its advantages and disadvantages are associated with design of itself. Therefore, they are widely used. Design model "*D*" of horizontal turn plow's moldboard provides smooth plowing, its cylindrical surface is simply for manufacturing. However, its surface worse controls layer trajectory and poorly turnover it. Information about their use in conditions of Uzbekistan is not available. Design model "*E*" of moldboard, although not widely used, its main advantage is multifunctionality, as replacement of its wings allows us to adopt it to different agricultural conditions. However, it is intended for one side operation. Design model "*F*" of horizontal turn plow's moldboard has advantages such as multifunctionality and double-acting possibility, but it is very complexity and material quantity [22].

Visualization of qualitative assessment and analysis of characteristics in this way allows us to choose direction of design modeling on advantages, by combination of design models. However, although the development of a model based on the principle of "*Concept selection*" is initial stage of design and is subject to further development, it reduces above-mentioned design costs.

Next stage of development will be produce by design models that took 1–3 places on the rating of evaluation criteria. Variants of proposed design models "*G1*", "*G2*", "*G3*" of moldboards (**Table 6**), taking into account advantages of considered design models, have geometrically combined working surface, they consist separate parts and they have double-acting opportunity. On base of improved models analyzing and in results of studies, finally we can offer new geometric modeling direction (proposed direction "*G*") for improving models (**Figure 10**).

Proposed model G1 by B and D	Proposed model G2 by B and E	Proposed model G3 by B and F
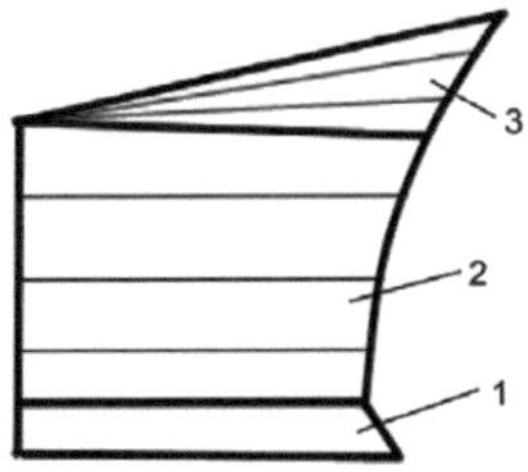	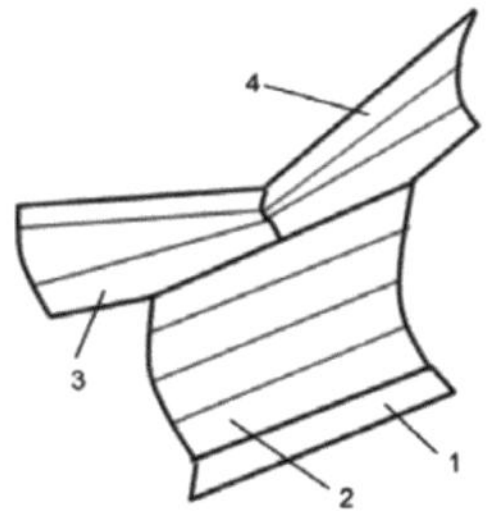	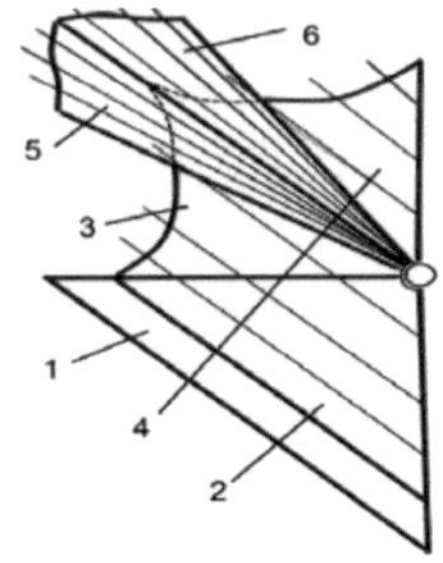
Plow for one side turning. Moldboard type–partial body. Working surface—geometrical combined. Plowshare (1), breast (2), wing (3).	For horizontal turn plow. Moldboard type–partial body. Working surface—geometrical combined. Plowshare (1), breast (2), right (3) and left (4) wings.	For horizontal turn plow. Moldboard type–partial body. Working surface—geometrical combined. Plowshare (1), main breast (2), right (3) and left (4) ancillary breast, right (5) and left (6) wings.

Table 6.
Description the design of proposed moldboard's models.

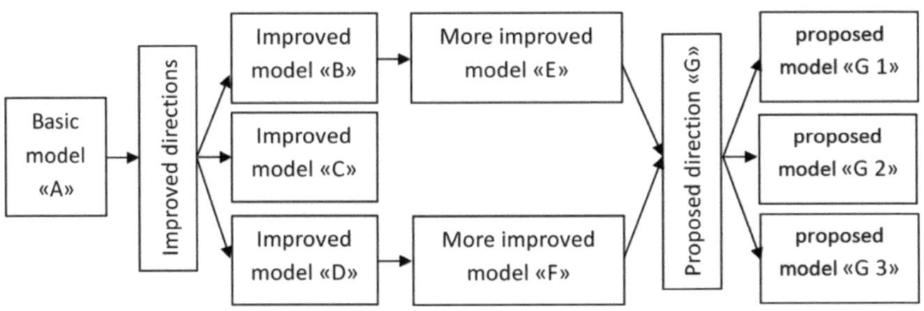

Figure 10.
Improving and proposing the moldboard designing directions.

4. Determining the integrative role of geometric modeling in tools' PLM for manufacturing

Modern manufacturing is base on CALS-technology (Continuous Acquisition and Lifecycle Support) or PLM-technology (information support of the product lifecycle management processes) by use the information science and communication technologies. PLM is an approach to design engineering and manufacturing high-tech and scientifically based product, concluding in use the information science and computer technology on all stages of the product lifecycle [23].

This aspect is actual in condition of developing countries like Uzbekistan, where using these technologies in manufacturing is considering as innovative process. One of the problems of this process is adaptation of these technologies on manufacturing, that is, transfer of engineering data into the PLM system, by integrating it with CAD/CAE/CAM systems, using the product's engineering database on base of PDM-technology (product data management).

The product's engineering data can be divided into three groups: constructive, functional, and technological. Let us consider the constructive data, which can be call also geometric data, that is necessary for the integration CAD and PDM

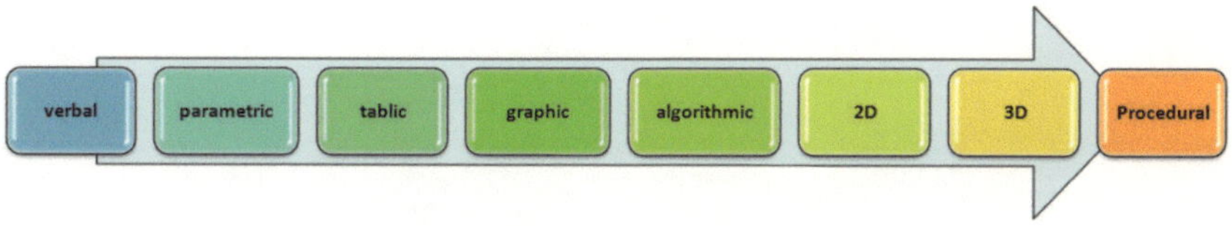

Figure 11.
Types and levels of geometric data information for design engineering.

systems. The product's geometric data is used not only in the company where it is produced, but also at all stages of product lifecycle from designing to manufacturing and post-manufacturing. Therefore creating the geometric database, which consist from the using different forms of geometric data (**Figure 11**), is very important in product lifecycle.

Product lifecycle includes period from origin necessity for creating the product up to its liquidations in consequence exhaustion of consumer characteristic. Primary stages of product lifecycle are selecting four main stages: designing, manufacturing, technical exploiting, and utilizing.

Though lifecycles of old and new product always formed unceasing cycle, because of not bright images, traditional lifecycles of each product were considered separately; whose initial stage is a designing but final salvaging. However, author founding on his conducting researches, offers to consider that beginning of PLM from creation an instrument from stone, bones, and wood by primitive man. Today someone cannot reject that base of modern industrial robot is an instruments of the stone age, so the end of "old" product is a beginning of "new" product. The present production conditions, in which production design steel is playing one of solving roles, relationship between "old" and "new" products in their lifecycle become reveals itself all more brighter. Coming from author offers separate stage of the designing on two: conceptual and engineering design. The conceptual design stage is based on the geometric modeling and it is closing stage of the product lifecycle having causal relationship between "post-manufacturing" (maintenance-utilizing) and "designing" stages. In current manufacturing conditions, geometric modeling has become the primary method and facility of the designing. At this stage, the product will designed on base of the relationships between exhaustion of consumer properties of an old product and necessity to creation a new (innovative) product.

The need for geometric data is at all stages of the life cycle of the product, especially at the initial stage, at the stage of "conceptual design". The Geometric Database created at this stage is directly or indirectly applied and at the subsequent stages of the product lifecycle, by integration CAD and PDM systems. It is necessary to note requirement to create "new product" basically it is formed in maintenance step of "old" product. Because at this stage, it is not only the Geometric data of "old" product in maintenance but also arises Geometric Data of "new" product in designing.

For example, let us consider the creation of Geometric Database in agriculture engineering tools manufacturing, which are necessary for enterprises participating in their products lifecycle [19, 23–26]. Creating this Database requires review, classification, and analysis of relevant information about agricultural machinery tools from geometric standpoint. This will enable us to identify the general and individual geometric features of these tools which will assists all participants in the lifecycle of data management in this process (**Figure 12**). The author is currently conducting research on the development of theoretical foundations and practical aspects of geometric modeling of agricultural machinery tools. Based on the results of the research models, algorithms and methods of designing these tools with moldboard surface by geometric modeling have been developed.

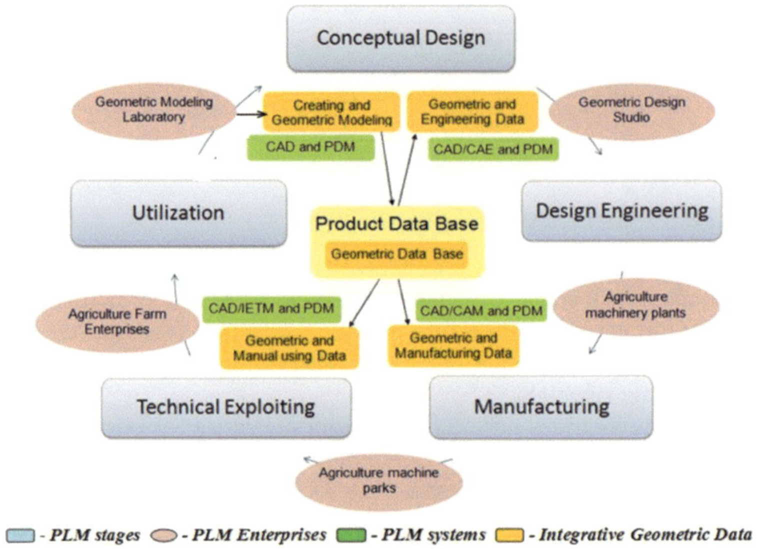

Figure 12.
Integration role of geometric modeling methods and systems in PLM.

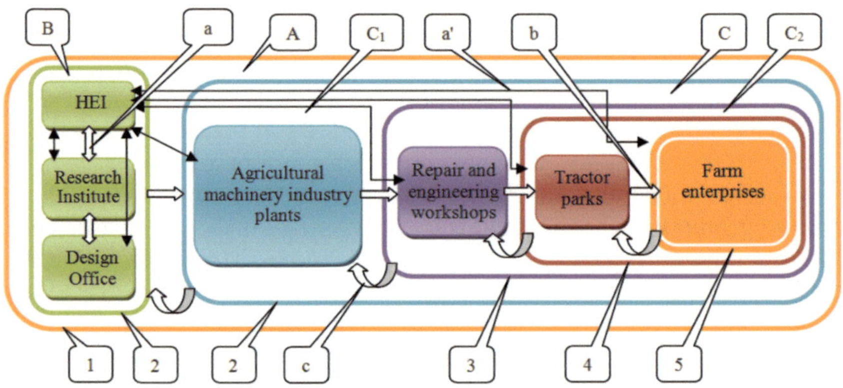

Figure 13.
Design engineering cluster's framework in agriculture.

As a result, sufficient information was collected to create a geometric database in different forms. It is possible to allocate following forms of the geometrical (graphic) data (information): verbal, graphical, parametric, algorithmic, 2D and 3D model, procedural.

Procedural data as manual using data can include all forms of geometric data. On base of research results, author worked out project-model of "Innovation cluster for design engineering and manufacturing in Agriculture engineering," when geometric modeling methods, applications, and systems play one of basic role (**Figure 13**). Framework of innovative cluster developed by system analyze of design engineering and manufacturing in Agriculture machinery tools PLM on view point of geometric modeling. In this cluster:

1. System framework: Over system ***A***–*"Government"*; Systems ***B***–*"University"* and ***C***–*"Industry"*; Subsystems ***C1***–*"Producer"* and ***C2***–*"Customer"*; Offered subsystem ***B'***–*"Innovative cluster on designing in agricultural machinery industry"*;

2. System components: ***B1***–*"Jurisdictional HEI (Higher Educational Institutions)"*, ***B2***–*"Jurisdictional Research Institutes"*, ***B3***–*"Jurisdictional Design Offices"*, ***C1***–*"Agricultural machinery industry plants"*, ***C21***–*"Repair and engineering workshops"*, ***C22***–*"Tractor parks"*, ***C23***–*"Farm enterprises"*. Offered component ***B1′***–*"Design office on geometric modeling"*;

3. System levels: ***1***-*National*, ***2***-*Regional*, ***3***-*Sub regional*, ***4***-*District*, *and* ***5***-*Sub district*;

4. System connections between components along cluster ***B′***: ***a***–inside system connections along designing the project, ***b***–consecutive outside (***output***) connections for implementation the project, ***c***–reverse outside (***input***) connections for correction the project;

5. ***a′***- functional connections of component ***B1′*** with other components along cluster ***B′***;

5. Conclusions

The proposed constructive geometric model of moldboard's working surface, allows to develop the multifunctional tools applicable in agricultural engineering, road building, mining, municipal service, and others branches of machinery. Parameterization of the moldboard's working surface facilitates the designer's work, expands the options for choosing the under developing moldboard's working surface and allows effectively to solve the constructive problems. The integration role of geometric modeling methods, systems, and applications allows efficiently apply them not only in design engineering process, and also in manufacturing processes of technical means. Creating the product's geometric database by CAD technologies became one of the necessary tasks of manufacturing, particularly engineering products. In contemporary conditions of using CALS technologies, "conceptual design" stage of innovative product by methods and facilities of geometric modeling is defining stage of the product lifecycle. So this application has signification in PLM, because geometric data will apply in all stages of PLM by geometric modeling methods, applications, and systems. The visualization of design-project process allows to develop the new production according to designing, manufacturing, and maintenance criterions. Effective use of these methods allows to reduce terms, labor, and material expenses of design process of new product. All figures and tables are produced by the author. All chapter materials are results of author researches, conducting in doctoral studies period by sponsorships of Government of the Republic of Uzbekistan in the Tashkent Institute of Irrigation and Agriculture Mechanization Engineers and Bukhara Institute of Engineering Technology [27].

Author details

Tojiddin Juraev Khayrullaevich[1*], Murodov Nusrat Murtazoyevich[2]
and Naimov Sandjar Tulkunovich[3]

1 Innovations and Scientific-Pedagogic Personals Training Department, Tashkent Institute of Irrigation and Agricultural Mechanization Engineers, Bukhara, Uzbekistan

2 Tashkent Institute of Irrigation and Agricultural Mechanization Engineers, Bukhara, Uzbekistan

3 Department of Descriptive Geometry and Engineering Drawing, Bukhara Engineering-Technological Institute, Bukhara, Uzbekistan

*Address all correspondence to: tojiddin_1968@mail.ru; tojiddin_1968@tiiamebb.uz

References

[1] Edict of the President of the Republic of Uzbekistan. About strategies actions on the further development of the Republic of Uzbekistan. Tashkent. Newspaper "Xalq so'zi", No. 28 (6692), February 8, 2017

[2] Production design. Principles, tools and techniques. ME 1007 design principles. Concept selection. Available from: http://www.product_design_principles.pdf

[3] Tyalve E. Short Course of the Industrial Design. Translation by Kunin PA. Moscow: Mashinostroenie; 1984. 192 p

[4] Juraev TX. Geometrical Modeling Principles of Meliorative and Agriculture Machines Tools. Monograph. Saarbrucken: Lambert Academic Publishing; 2015. ISBN 978-3-659-66832-6. Available from: http://dnb.d-nb.de

[5] Voloshinov DV. Theory of automation of design of objects and processes on the basis of methods of constructive geometric modeling: Annotation of Dissertation: SPb.: SPbSPU, 2010. 33 p

[6] Goryachkin VP. Collected Works. Vol. 1–2. Moscow: Kolos; 1965 648 p

[7] Korabelski VI. Justification of the shape and parameters tillage bodies using geometrical modeling of the main technological requirements [thesis of dissertation]. Chelyabinsk: ChIMEA; 1988. 450 p

[8] Gyachev LV. Theory of Share-Moldboard Surface. Zernograd: Gosgortekhizdat; 1961. 317 p

[9] Shestopalov KK. Machines for Earthworks: School-Book. Moscow: MADI; 2011. 145 p

[10] Bosoy EC et al. The Theory, Design and Engineering of the Agricultural Machines. Moscow: Mashinostroenie; 1978. 568 p

[11] Mould boards. Available from: http://www.max-agro.ru

[12] Catalogue of special machines. ISP GROUP. Available from: www.isp-group.ru

[13] Voloshinov DV. The Theory to automations of the designing object and processes on base of the methods of constructive geometric modeling [abstract of DSc thesis] Saint-Petersburg; 2010

[14] Biliev AA. The motivation of the technological scheme of the plow for smooth two-tier plowing [abstract of PhD thesis]. Moscow. VIM; 1992. 22 p

[15] Dubanov AA. Methods and algorithms to approximations of the technical surfaces turning round [abstract of PhD thesis]. Moscow. MUPP; 1997

[16] Murodov SK et al. Descriptive Geometry. School-Book. Tashkent: IQTISOD-MOLIYA; 2008

[17] Voloshinov DV. Constructive Geometric Modeling. Theory, Application, Automatization: Monograph. Saarbrucken: Lambert Academic Publishing; 2010. 355 p

[18] Kobec AS, Sokol SP, Korabelski VI, Naumenko NN, Kobec AN. Design-project shape geometry of formative alternating curved surface of bracket. Available from: http://www.nbuv.gov.ua

[19] Juraev TX. Creating the Geometric Database for Product Lifecycle Management System in Agricultural Engineering. International Conference on Information Science and Communications Technologies ICISCT

2017 Applications, Trends and Opportunities. 2–4 November 2017; TUIT, Tashkent, Uzbekistan. IEEE Catalog Part Number: CFP17H74-CDR, ISBN: 978-1-5386-2167-7. Available from: https://www.researchgate.net/publication/321821311

[20] Juraev TX. Conceptual designing of mould board's surface by geometrical modeling. American Journal of Mechanics and Applications. 2017;**5**(4): 28-33. DOI: 10.11648/j.ajma.20170504.11. ISSN: 2376-6115 (Print); ISSN: 2376-6131 (Online). Available from: http://www.ajmechanics.org/archive/621/6210504

[21] Flat plowing plow. SU1732826 (A1)-1992-05-15 Available from: http://www.Espasenet

[22] Juraev TX. Decision maintenance management problems in agriculture engineering by constructive geometric modeling methods. Maintenance Management. DOI: 10.5772 / IntechOpen.81969

[23] Jayakiran Reddy E, Venkatachalapathia N, Pandu Rangadu V. Development of an approach for knowledge-based system for CAD modeling. Selection and/or peer-review under responsibility of international conference on materials manufacturing and modeling (ICMMM-2017). Science direct. Materials Today: Proceedings. 2018;**5**:13375-13382

[24] Juraev TX. Geometric modeling for organizing innovative cluster on designing in agricultural machinery industry. Proceedings of the international conference on integrated innovative development of Zarafshan region: achievements, challenges and prospects. Vol. II. 26–27 October, 2017. Navoi, Uzbekistan: NSMI. pp. 248-251

[25] Juraev TX. Geometrical modeling of agricultural machines as a mean of engineering provision of agriculture production. The collection materials of VIII-th International scientific conference "Modern technologies in agricultural production and education." Kemerovo, Russia: KSIA. 2017. Available from: http://www.ksai.ru/upload/files/sborniki/inyaz_2017/files/assets/basic-html/page119.html

[26] Matejun M. The process of opportunities exploration and exploitation in the development of SMEs' innovativeness. Management and Production Engineering Review. 2018; **9**(3):3-15. DOI: 10.24425/119529

[27] Juraev TX. Geometric modeling of agriculture and meliorative machines' tools [abstract of PhD dissertation]. Tashkent University of Information Technologies; 2019. Available from: www.tuit.uz

Chapter 13

Manufacturing a Ceramic Water Filter Press for Use in Nigeria

Ebele A. Erhuanga, Isah Bolaji Kashim, Tolulope L. Akinbogun, Olusegun A. Fatuyi, Isiaka A. Amoo and Daniel J. Arotupin

Abstract

A significant proportion of Nigerian households lack access to improved and safe drinking water supplies. This has resulted in high incidences of diarrhoeal-related deaths in the country, especially among young children. Several studies have shown that point-of-use water treatment options such as ceramic filtration are effective in reducing the occurrence of water-borne diseases; however, its use in Nigeria has been significantly low. There is a need to build entrepreneurial capacity among local potters and potteries to drive the scale up of ceramic water filter production across the nation in order to create demand for the filters, seeing that huge potential for its sales abounds. However, the high cost of acquisition of the ceramic water filter press, which is the most essential equipment in the production of the water filters, is a major limitation to the scale up of ceramic water filter production in the country. The goal of the study was to manufacture a ceramic water filter press, by adapting an existing design, using locally sourced materials and manpower, to achieve lower cost. The resulting filter press cost approximately $1000, proving the viability and cost efficiency of the local manufacture of ceramic water filter presses in Nigeria.

Keywords: ceramic water filters, filter press design, household water treatment, manufacturing, Nigeria

1. Introduction

Water is most essential for sustaining life and enhancing the quality of life, but it can transmit diseases. When adequate access to clean, safe water is lacking, incidences of waterborne diseases become rampant [1, 2]. Unsafe drinking water is one of the major causes of diarrhoeal diseases, which are known to be a leading cause of mortality globally especially in children aged five and below [3]. The 2015 WHO/UNICEF Joint Monitoring Programme (JMP) update reports that 69% of Nigeria's population use improved drinking water sources, which are presumed to be safe [4]. However, due to non-functionality, unsustainability, and lack of proper maintenance of most improved water sources, they are often of non-satisfactory quality [1, 5]. Therefore, the reality is that a lesser percentage of Nigerians than presented actually have access to safe drinking water. Furthermore, even where there is access to safe water, because most of these water sources are not located on premises or piped directly into the houses, there is the risk of contamination in the process of collection, transportation, and storage, thereby leaving the initially safe water unsafe at the point of consumption [2, 6]. It is therefore essential to ensure water is safe for drinking at the point of consumption. Point-of-use water treatment implies any water treatment system that purifies water at the point of consumption and it involves effective treatment and safe storage. It has been identified as an important public health intervention which serves to reduce the faecal-oral transmission of diarrhoeal diseases [7].

Recent studies on point-of-use household water treatment systems, suggest that ceramic water filters are the most sustainable and lowest cost options for water purification in developing countries [8]. The essential raw materials, basically clay and combustible bio-wastes, required to make this technology available and accessible in Nigeria are locally available in large quantities. However, there is a wide knowledge gap in the exploration and development of the technology of manufacturing ceramic water filters in the country. As much as there exists a need for household water treatment method such as the ceramic water filters, not many manufacturers engage in the production of ceramic filters. The springing forth of many peri-urban settlements in many Nigerian cities like Akure leaves the nation fraught with an urgent need to explore innovative solutions to put an end in sight to the prevalent water-related health challenges.

While household water treatment and safe storage systems have been considered as effective, low-cost alternatives and a reliable means of achieving safe water at point of use, having shown to significantly reduce diarrhoeal prevalence [6, 7, 9]; very few potters engage in the making of the ceramic water filters. In Nigeria, there are two factories that currently produce ceramic water filters, although production is fraught with many challenges such as understanding the technology behind the working of the filtration system. The major challenge however, to the establishment of a ceramic water filter production facility is the acquisition of the filter press machine.

The ceramic filter press machine is the priority piece of equipment required in the production process of ceramic water filters [10, 11]. The filter press machine, which is mostly hydraulic operated, is used to form the filters into its shape by the application of pressure to the clay mixture in-between a set of moulds. This method of forming is most suitable for making ceramic water filters because a non-plastic material mix is desired and therefore can be only formed successfully by semi-dry pressing techniques. This all-important equipment for the production of ceramic water filters is quite expensive to purchase, with very high shipping and importation costs and tariffs.

Personal communications in a pilot study with operators of ceramic water filter factories in Nigeria reveals that the cost of acquisition of a piece of filter press machine with its corresponding aluminium moulds ranged from $3000 to $3500 (USD). This is also confirmed by other researchers [10], stating that the cost of this press is estimated at over $3000 and therefore is considered a fundamental limiting factor to production of ceramic water filters to meet demands in areas where it is needed. While the Resource Development International - Cambodia (RDIC) approximated the cost at $2300, excluding shipping and handling costs [12]. This is too high an investment cost for a start-up ceramic/pottery business to bear considering the economic conditions in the country. Therefore the only feasible option to the making of ceramic water filters in Nigeria, to improve access to safe drinking water at the point-of-use, is to resort to the design and fabrication of a filter press machine using locally available materials.

The Potters Without borders (PWB) is one of the organizations that have carried out research on ceramic water filters and design of hydraulic filter press machine [10].

The PWB filter press machine design was adopted for this study, whose objective was to design and fabricate a hydraulic filter press unit using locally sourced materials with a view to promote the affordability and availability of this technology for the manufacture of ceramic water filters, consequently increasing access to safe drinking water in Nigeria.

2. Previous works on ceramic water filter press design

At its inception by Fernando Mazariegos, the ceramic pot water filter was shaped by hand on the potters' wheel. But in the 1980s, the Central American Institute of Industrial Research and Technology (ICAITI) introduced the use of hydraulic presses in the shaping of ceramic water filters resulting in more efficient ceramic water filter production and performance [10]. However, other literature [13] reports that the first press and the first set of moulds were developed to standardize the shape of the ceramic water filter (see **Figure 1**).

While the Potters Without Borders (PWB) press design is the most commonly used, other attempts have been made to explore different press designs to improve the workings and efficiency of the presses in the production of ceramic water filters and to meet the specific socio-economic needs of varying localities. The PWB filter press design operates with a 20-ton hydraulic jack and a hand lever for lifting and lowering the H-slide to which the male mould is attached. It produces the flat-bottomed ceramic water filters, using a set of aluminium moulds.

A recent study [14] on a multi-component water treatment, reported that they created a simple plastic press mould to shape the ceramic component of their water filtration system with the aim to improve efficiency and allow for easy replication. (see **Figure 2**).

Another study [10] designed a low-cost filter press with the goal of less than $200 in cost, less manpower requirement and shorter manufacture time. Their work concentrated on designing and prototyping a low-cost, filter press using

Figure 1.
Ron Rivera working on the first ceramic filter press [13].

Figure 2.
Modelled diagram of the press mould and product [14].

locally-sourced materials. They attempted to achieve a lower filter formation pressure as a key requirement to reducing the cost, considering that using a 2-ton car jack instead of the 20-ton hydraulic jack used by PWB would greatly reduce cost. The press was designed for the round-bottom filters and adopted an inverted design in which the car jack was mounted to the frame headstock and the female mould was suspended on the underside of the jack elevator while the male mould sat on the base [10]. For the moulds, they improvised with the use of inexpensive aluminium bowls (see **Figure 3**).

A group of researchers [11] in their study described the use of a 30-ton manually operated hydraulic press developed and manufactured by MEC Ltd., India. The press makes use of a screw system to lower and lift the male mould which is attached to the die screw connector plate, while the female mould sits on a base plate which is attached to the hydraulic jack (see **Figure 4**). This press produces the flat-bottomed, frustum shaped filters of 23 cm height with 25.5 cm base diameter.

The Ceramic Filter Manufacturing Manual [15] developed by Pure Home Water, reported two types of press designs for shaping ceramic water filters; the Potters for Peace (PfP) press and the Mani press. The PfP press design as described in the text is a portable press that uses a 20-ton hydraulic jack with a removable female mould while the male mould is attached to a moveable shaft on the frame. It operates a crankshaft system which allows for the lifting and lowering of the shaft that holds

Figure 3.
A low-cost filter press prototype [10].

Figure 4.
Filter press operated with screw and hydraulic system [11].

Figure 5.
Operating the portable PfP press with crank system [15].

the male mould. The hydraulic jack is positioned above the male mould after it has been lowered into the female mould which contains the clay (see **Figure 5**).

The Mani press has both its male and female moulds attached; while the male mould is attached to an extendable table, the female mould is attached to the press

Figure 6.
The Mani press [15].

frame. It uses an 8-ton hydraulic jack and works with a pulley system that operates with a hand crank for lifting and lowering the female mould (see **Figure 6**).

The mould in the PfP press described in the Ceramic Filter Manufacturing Manual [15] is made of nylon while the material used to make the Mani press moulds was not stated in their report but can be made of concrete or metal. It, however, concluded that the Mani press delivered greater advantage and ease in use than the portable PfP press. The RDIC manual [12] describes a fully automated hydraulic system-operated ceramic water filter press. It uses a set of metal moulds, most likely aluminium. The male mould is attached to the frame headstock while the female mould is attached to a moveable shaft which is controlled by the hydraulic system which works with the use of an electric motor (see **Figure 7**). This action controls the press and the release of the clay filter mix in between the moulds.

The features of the various designs of ceramic water filter presses reviewed in the course of this study are presented in **Table 1**.

After a review of the designs of ceramic water filter presses as discussed hitherto, the PWB ceramic water filter press design was adopted based on the following considerations:

A non-electrically operated press was desired to overcome the challenge of poor electricity supply within the country;

Figure 7.
An electric motor driven hydraulic press [12].

Ref. no	Description	Type of filters	Position of moulds	Mould material	Mould moving mechanism	Mode of operation
14	Hand press mould	Ceramic filter component	—	Plastic	—	Hand/ manual
10	Low-cost filter press	Round bottom	Inverted; female above	Aluminium bowls	—	2-ton hydraulic car jack
11	MEC India manufactured press	Flat bottom	Upright; male above	—	Hand-operated screw system	30-ton hydraulic jack
15	PfP portable press	Flat bottom	Upright; removable female mould positioned below	Nylon	Hand-operated crank system	20-ton hydraulic jack
15	Mani press	Round bottom	Inverted; male attached to extendable surface	—	Crank-operated pulley system	8-ton hydraulic jack
12	RDI-C	Flat bottom	Upright; male above	Metal	—	Automated hydraulic system
10	PWB	Flat bottom	Upright; male above	Aluminium	Hand-operated lever	20-ton hydraulic jack

Table 1.
Features of the various types of ceramic water filter presses reviewed in this study.

The use of a fully manual system was also not desirable because it will increase the time taken to press one filter; therefore a hydraulic press mechanism was desired;

A lever was preferred for the lowering and lifting of the moulds, to the crank (as in the portable PfP press [15]) and the screw (as in [11]) because it makes the filter pressing more cumbersome and time consuming;

The moulds were preferred fitted to the press frame to overcome the challenge of misalignment of moulds, possible in removable moulds, and as well, the inconvenience and health hazard of lifting heavy moulds in each process of filter pressing (as in the portable PfP press [15]).

Based on these specific requirements, the Potters Without Borders (PWB) ceramic water filter press design was adapted for manufacture in Akure, Nigeria. The PWB ceramic water filter press is said to have several benefits with respect to design and operation. Its high-strength (20-ton) design allows the pressing of flat-bottom filters [10] while creating stability and preventing deformation in the shaped filters. The flat-bottom filters are said to provide more surface area and therefore higher flow rates [10]. Some of the adjustments made to the PWB filter press design, included the replacement of the hydraulic car jack with a locally fabricated industrial hydraulic jack, as well as the design and manufacture of the press mould to fit locally available wide-rimmed plastic containers to meet the water needs in larger households.

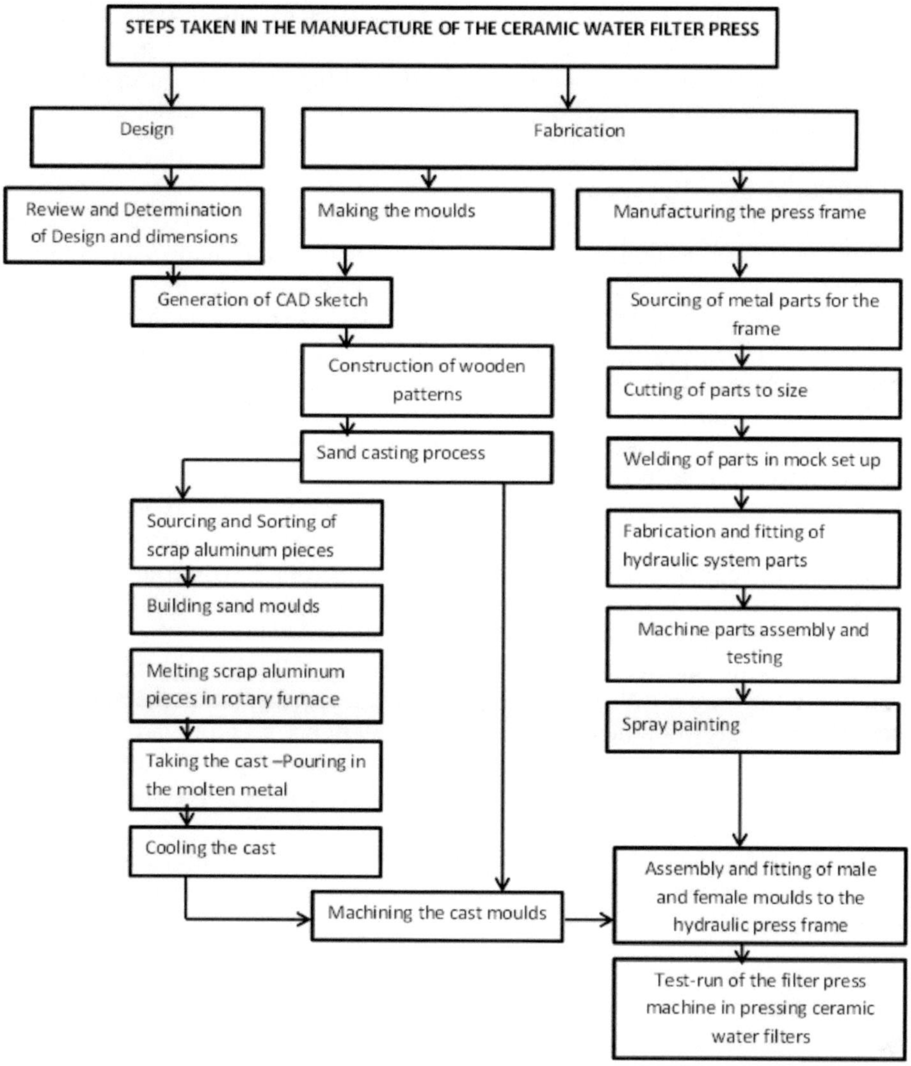

Flow chart of steps taken in fabricating the ceramic water filter press.

3. Fabricating the hydraulic press machine

It became expedient to fabricate a hydraulic press machine to facilitate the shaping of the ceramic water filters by the press cast method. This is the most suitable method of forming the ceramic filters because the mix is highly non-plastic and hence cannot withstand other ceramic forming techniques besides slip casting which is not very feasible at the desired dimensions of ceramic water filters.

For this study to ensure the economic feasibility, sustainability and hence the scalability of the manufacture of the ceramic water filter press in Nigeria, it was important to set a cost limit for fabricating the press; and this was set at 350000 naira (approximately $1000). This was done considering the issue of low access to capital for start-ups, which is common in the country. This study, however, intends to encourage local potters to venture into the production of ceramic water filters by alleviating some of the cost-related challenges of setting up a filter production unit.

All the materials and manpower used in fabricating this press were sourced from within the country. The hydraulic press machine typically consists of two parts; the moulds and the frame which holds the moulds and the hydraulic component. The procedures engaged in the making of both parts are discussed further.

3.1 Making of the filter moulds

The mould for the filter press machine was designed and made using aluminium as material, which was shaped using the sand casting method. The processes involved in the making of the filter mould include; generating a CAD drawing (see **Figure 8**), detailing the dimensions of the moulds; and the making of a wooden mould patterns (see **Figures 9** and **10**) from which sand moulds were derived.

The mould design was generated during the course of the study using dimensions which were estimated by the researcher to produce a ceramic water filter that would fit into commonly available wide-rimmed large plastic containers. The size of the container was used as mark up for the determination of the dimensions of the moulds. The core and drag mould components were designed to give a pressed ceramic filter product of 30 mm thickness all round; this is to accommodate the high shrinkage possible in most plastic ball clays available for use in South West Nigeria; as well as to allow for longer contact time with silver for the inactivation of pathogens in water and greater possibility of trapping the pathogens as they travel through the filter walls. With this design sketch, a wooden pattern made of cut out pieces of 2-inch plywood held together with resin bond, was derived. The pattern is highly essential to the process because the sand moulds which was used for casting the metal form is taken from it. So it is important to ensure correctness of dimensions and form in the wooden pattern.

The process of making of the sand moulds included filling up firmly, a square-shaped wooden frame in which the wooden pattern has been placed with fine sand (see **Figure 11**); after which the pattern is taken out and the sand is smoothened out

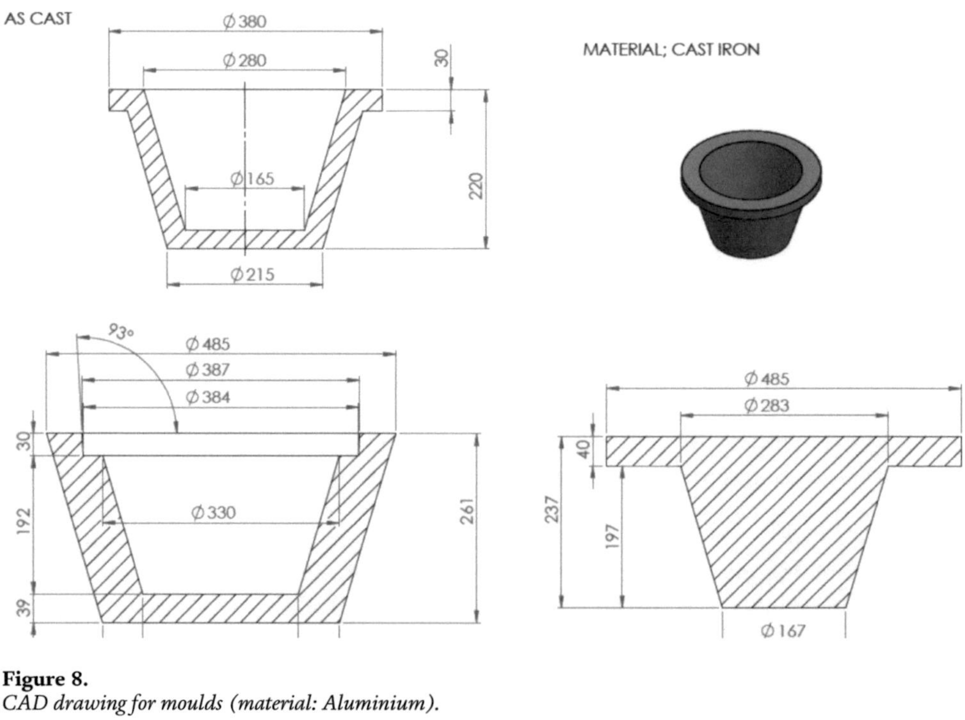

Figure 8.
CAD drawing for moulds (material: Aluminium).

Figure 9.
Wooden patterns for the mould.

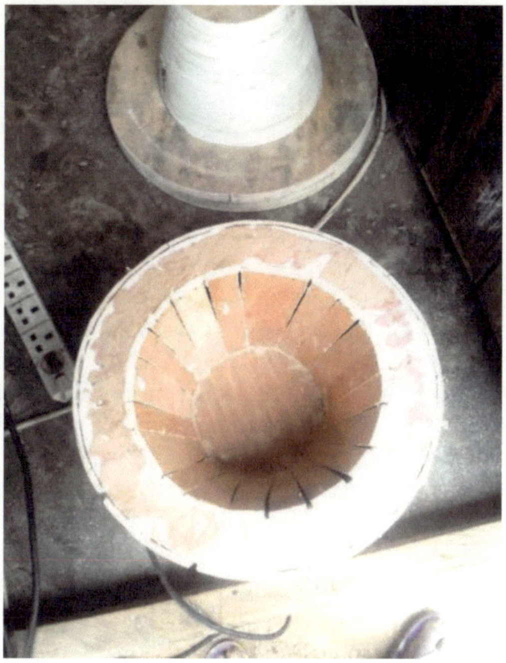

Figure 10.
Top view of wooden patterns for the mould.

using a metal spoon (see **Figures 12–14**). The metal cast was then taken from the prepared sand mould.

Pieces of waste aluminium collected from the local scrap market were charged into the rotary furnace and melted (see **Figure 15**) at temperatures between 600 and 700°C. The crucible bearing the molten aluminium was removed from the

Figure 11.
Filling the frame with sand.

Figure 12.
Pattern taken out.

furnace using a pair of furnace tongs (see **Figure 16**) and the crucible holding the molten metal was set in a 2-man carrier rod (see **Figure 17**).

It is important to remove dross and check for unmolten particles of other metals before casting (see **Figure 18**). The molten metal is then poured into the sand moulds by means of crucible tongs and carrier rod (see **Figures 19** and **20**).

In the process of pouring in the molten material, it is important to poke at it using a metal rod to aid the removal of any air bubbles that may have been trapped in while pouring (see **Figure 21**). The metal cast is afterwards left to cool for about 24 hours before it is removed from the mould (see **Figure 22**). The surface finish of the cast aluminium mould is mostly dull, lacks lustre and sometimes presents tiny holes as seen in **Figure 23**. Polishing the metal is therefore important to give a more usable finish to the cast aluminium moulds (see **Figure 24**).

The last phase in the making of the mould was the machining and polishing of the cast. Aluminium was the material used to make the moulds in this study. This is because aluminium is a non-rust metal and it is more affordable than stainless steel and can easily be machined because it is a relatively soft metal. Aluminium is also a very available material in most scrap markets across the country, and hence easy to access for this purpose. The machining or polishing of the moulds was carried out using a horizontal lathe machine in a privately-owned engineering workshop.

Figure 13.
Smoothening the sand mould.

Figure 14.
Finished sand mould.

Figure 15.
Process of melting the scrap aluminium in a rotary furnace.

Figure 16.
Removing molten aluminium from the furnace using a pair of tongs.

Figure 17.
Crucible set in the carrier rod in readiness for casting.

Figure 18.
Stoking the molten metal to remove dross and other particles.

Figure 19.
Pouring in the molten metal into the sand mould using furnace tongs.

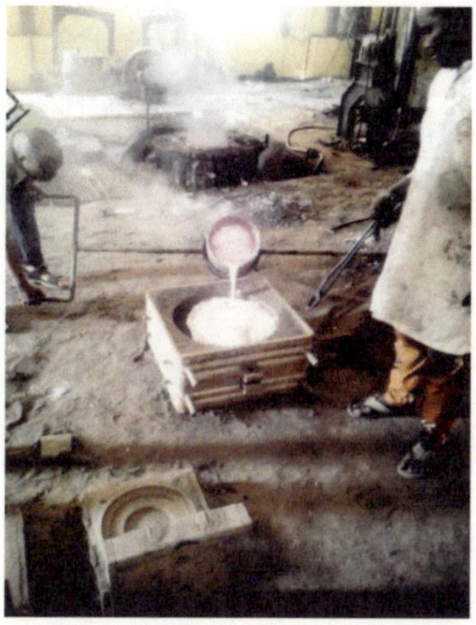

Figure 20.
Casting process using the crucible carrier.

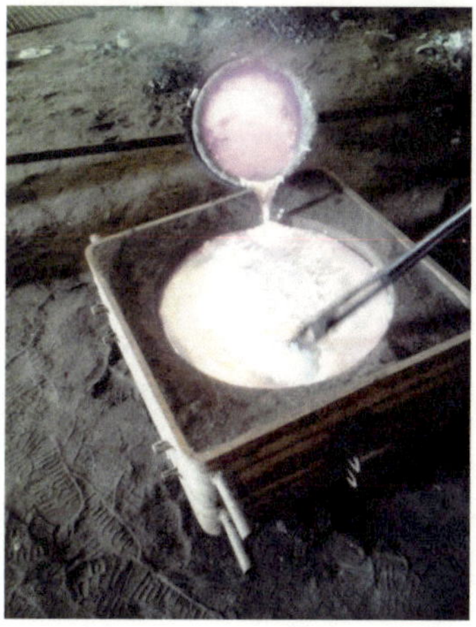

Figure 21.
Poking the poured-in metal to remove trapped air.

Figure 22.
Cooling.

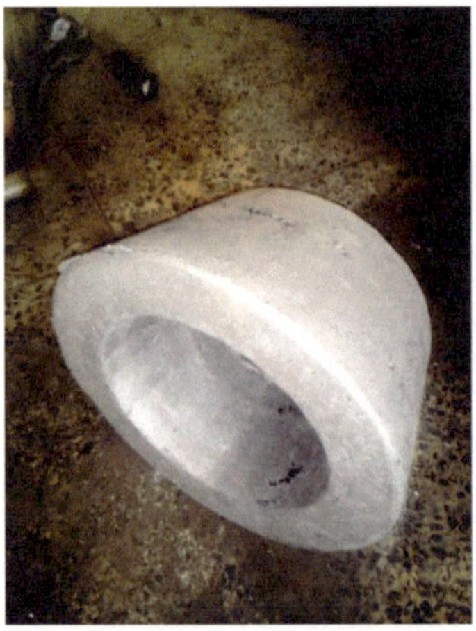

Figure 23.
Cast aluminium moulds.

Figure 24.
Polished aluminium mould.

3.2 Making of the hydraulic press frame

The frame of the hydraulic press machine was made from cast iron and steel parts. The design for the frame was adapted from the Potters Without borders (PWB) ceramic water filter press design (see **Figure 25**). The PWB filter press design incorporates the use of a removable car jack as its hydraulic mechanism. The design for this study has incorporated a hydraulic controller system which is comprised of a box, an industrial jack to drive the pressing mechanism which is expected to be more durable than the car jack over time and continued use; and a pressure gauge to measure the pressure applied in the pressing of each filter to enhance consistency in production.

The metal parts for the frame were sourced from Akure and Ibadan in Southwest Nigeria. Cast iron was the major material from which the parts of the frame were made. Some parts were also of made of steel. The long metal parts were cut into dimensions (see **Figures 26** and **27**) and holes were drilled through them to enable assembly of the frame using nuts and bolts. Bolting was preferred to welding in the assembly of the machine parts, to allow room for adjustments and for easy movement and transportation of the machine. The cutting and welding of the frame

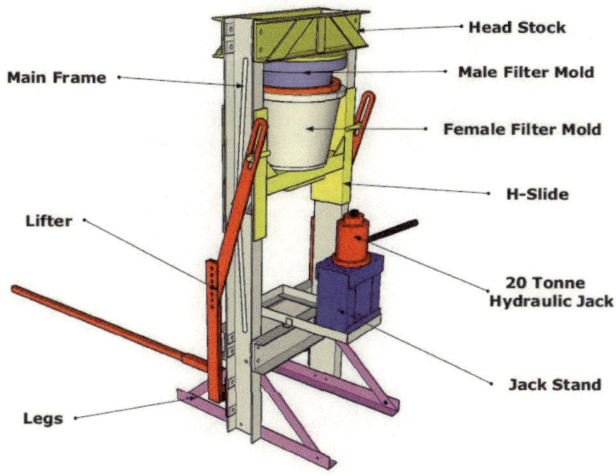

Figure 25.
PWB design of press machine [10].

Figure 26.
Cut out metal parts for the frame.

Figure 27.
Metal parts of frame in mock assembly.

Figure 28.
Press frame with moulds mounted.

Figure 29.
Installation of the lever mechanism.

Figure 30.
Testing the installed lever and jack.

Figure 31.
Press with hydraulic system installed.

Figure 32.
Finished ceramic water filter press.

was followed by the mounting of the moulds. The male component of the mould was bolted onto a metal plate which is welded to the headstock of the frame, and the female component was fitted via bolting onto the moveable H-slide (see **Figure 28**). The lever system which is used to control the lifting of the H-slide bearing the female mould during pressing and release of the moulds, was subsequently fixed in place (see **Figure 29**) and test run to assess the mould alignment (see **Figure 30**). The hydraulic jack was thereafter installed and tested in operation with the lever as shown in **Figure 31**. Finally, the hydraulic control box was installed and connected to the jack and the entire frame was sprayed with paint to improve its aesthetic and prevent rusting (see **Figure 32**). The making of the frame and the hydraulic control box, as well as the assembly of the moulds was done at Danzaki Engineering Services, a privately-owned mechanical engineering workshop in Akure, Nigeria.

4. Results and discussion

The outcome of the study showed the local availability of the required skills and material resources to locally manufacture a ceramic water filter hydraulic press machine in Nigeria. The total cost of the local production of the press though slightly above the set target, is approximated at $1000 USD and is about one-thirds of the cost of acquiring a press of similar specifications of foreign origin without the attending shipping and clearing costs.

The manufactured ceramic water filter press was effective in the shaping of ceramic water filters as indicated in the evenness in form and thickness of the filters pressed during a test run of the filter press (see **Figure 33**).

The technical specifications of the ceramic water filters produced from the manufactured filter press are outlined as having an inner height of 15 cm and inner diameter of 28.5 cm; with an estimated volume capacity of 12 L. This is specified to fit into a 30-L capacity bucket with a rim diameter of 30 cm. Shrinkage allowance of 10% was estimated and factored into the design to ensure the resulting filters fit onto the desired bucket.

However, there were a few limitations to the study as outlined thus: At the size required for the set of moulds, it was difficult to find a lathe machine of a size that could hold the cast moulds for machining. Therefore, alternative materials may be explored besides aluminium, especially such materials as would not require

Figure 33.
Freshly pressed ceramic water filter using the fabricated press.

machining/polishing. Also, there were issues surrounding the dimensions presented in the CAD sketch as generated by a draughtsman, this resulted in error in the moulds cast. This was, however, corrected by altering the dimensions of the mould during the process of machining in order to achieve even thickness around the product; and this action reduced the size of the mould and hence the resulting filter is shorter than other filters available.

5. Conclusions

This book chapter documents the procedure and results obtained in a study carried out to explore the local manufacturing of a ceramic filter press in order to prove the viability and cost efficiency of producing it locally as compared with the cost of acquiring the imported presses. This is in a view to encourage the set-up of more ceramic water filter producing factories in Nigeria, thereby bringing closer home the technology that would make clean, safe water more accessible and available to communities and households across the country.

The study indicates that ceramic water filter presses with hydraulic components as well as its corresponding set of moulds can be successfully and inexpensively manufactured in Nigeria, using all materials and skills sourced locally from within the country.

Acknowledgements

The authors would like to acknowledge the Management of the Federal University of Technology, Akure and TETFund for providing funding for this work under the IBR grant with reference number, VCPU/TETFund/155.

We appreciate Engr. A. Smart and Engr. Idowu of EMDI, Akure, for analyzing the possible designs for the hydraulic press system with us at the commencement of this study; Mr. Yekin Obe and staff of Foundry Department, FIIRO, Lagos, for their assistance with the casting of the filter moulds; and Mr. J. O. Oke and Mr. M. Familusi of the Industrial Design Department, FUTA for their assistance in the entire course of the study and specifically for test running the equipment after its manufacture. Our appreciation also goes to Robert Pillers for reviewing the filter press in progress and making useful inputs that led to some adjustments.

Conflict of interest

We would like to declare that there is no conflict of interest.

Author details

Ebele A. Erhuanga*, Isah Bolaji Kashim, Tolulope L. Akinbogun, Olusegun A. Fatuyi, Isiaka A. Amoo and Daniel J. Arotupin
Federal University of Technology, Akure, Nigeria

*Address all correspondence to: eaerhuanga@futa.edu.ng

References

[1] Dinka MO. Safe drinking water: Concepts, benefits, principles and standards. In: Glazan M, editor. Water Challenges of an Urbanizing World. Rijeka: IntechOpen; 2018

[2] Osiemo MM, Ogendi GM, M'Erimba C. Microbial quality of drinking water and prevalence of water-related diseases in Marigat Urban Centre, Kenya. Environmental Health Insights. 2019;**13**:1178630219836988. DOI: 10.1177.1178630219836988

[3] Centers for Disease Control and Prevention (CDC). Global Diarrhea Burden [Internet]. 2015. Available from: https://www.cdc.gov/healthywater. [Accessed: 27 September 2019]

[4] UNICEF and World Health Organization. Progress on Sanitation and Drinking Water–2015 Update and MDG Assessment; 2015. Available form: https://www.washmatters.wateraid.org

[5] Erhuanga E, Kashim IB, Akinbogun TL. Development of ceramic filters for household water treatment in Nigeria. Art Design Review. 2014;**2**(1):6-10

[6] Kleiman SL. Ceramic filter manufacturing in northern Ghana: water storage and quality Control [thesis]. Massachusetts Institute of Technology; 2011

[7] World Health Organization. Water quality interventions to prevent diarrhoea: Cost and cost-effectiveness. WHO/HSE/WSH/08.02. Geneva: Public Health and Environment, World Health Organization; 2008

[8] Brown J, Sobsey MD. Microbiological effectiveness of locally produced ceramic filters for drinking water treatment in Cambodia. Journal of Water and Health. 2010;**8**(1):1-10

[9] Baumgartner J, Murcott S, Ezzati M. Reconsidering 'appropriate technology': The effects of operating conditions on the bacterial removal performance of two household drinking-water filter systems. Environmental Research Letters. 2007;**2**:024003. UK: IOP Publishing Ltd.

[10] Henry M, Maley S, Mehta K. Designing a low-cost ceramic water filter press. International Journal for Service Learning in Engineering. 2013; **8**(1):62-77. Available from: https://sites.psu.edu/hese/2016/03/16/designing-a-low-cost-ceramic-water-filter-press/

[11] Gupta S, Satankar RK, Kaurwar A, Aravind U, Sharif M, Plappally A. Household production of ceramic water filters in Western Rajasthan, India. International Journal for Service Learning in Engineering, Humanitarian Engineering and Social Entrepreneurship. 2018;**13**(1):53-66. Available from: https://www.academia.edu/38980382/Household_Production_of_Ceramic_Water_Filters_in_Western_Rajasthan_India

[12] Hagan JM, Harley N, Hughes R, Chouhan A, Pointing D, Sampson M, et al. Resource Development International - Cambodia Ceramic Water Filter Handbook. Version 1.3. Cambodia: Phnom Penh; 2009

[13] Potters for Peace. Remembering Ron Rivera. [Internet] 2019. Available from: https://pottersforpeace.org/?page_id=645 [Accessed: 20 July 2019]

[14] Concannon M, Genga J, Jonwal P, Kumar A, Meena P, Nash P, et al. Designing a water filtration device to remove chemical and biological contamination in Mandi District [thesis]. Worcester Polytechnic Institute; 2018

[15] Miller TR, Watters TR. Pure home water ceramic filter manufacturing manual [thesis]. Cambridge, Massachusetts: Massachusetts Institute of Technology, Department of Civil and Environmental Engineering, Massachusetts; 2010